42 REASONS TO HATE THE UNIVERSE

AND
ONE REASON
NOT TO

CHRIS FERRIE
WADE DAVID FAIRCLOUGH
BYRNE LAGINESTRA

Published by Sourcebooks
P.O. Box 4410, Naperville, Illinois 60567-4410
(630) 961-3900
sourcebooks.com

Cataloging-in-Publication Data is on file with the Library of Congress.

Printed and bound in the United States of America.
KP 10 9 8 7 6 5 4 3 2 1

CONTENTS

PART VII: SO LET'S GO CRAZY

INTRODUCTION
Why hate the Universe?

Twinkle twinkle little star,
How I wonder what you are...

We were all sung this song as children as we gazed up at the heavens at the tiny points of light in the night sky in wonder and awe at the beauty and mystery of the Universe. Of course, today we wonder less about the stars. Science has given us a new lens on the Universe, allowing us to see beyond what only our eyes can tell us. We have so much more to learn about the Universe, which seems to be a place of infinite possibilities. It is a place of wonder. It is a place of mystery. It is a place of beauty, love, and hope. And although science has taken us on a journey beyond the stars, these twinkling specks of light are where it all began.

Your parents probably told you that the stars were big balls of burning gas—which is wrong—and that there are millions of other stars in the Universe, but *our* star, the Sun, is special—which it isn't—and that special star shines just for you because you too are special—which you aren't.

There *is* beauty in the world though, or so it would seem. Our brains surge with endorphins—the "feel-good" chemicals—when we see a glowing sunset, catch a glimpse of a shooting star, or receive our first kiss. Optimistically, we want to believe the Universe made these moments specially for us. Alas, it is an illusion. The Universe cares about you, but only inasmuch as it wants you dead. Once you take the rose-colored glasses off, you soon realize that the Universe really wants nothing to do with you, and often it is trying to kill you. Our ability to see the world for what it really is—the naked truth—is granted by the power of science. But what does science say exactly about our existence or, more relevant to this book, our journey toward death?

Consider that you are just a group of atoms structured in a specific way for just long enough that you can try to understand this thing we call existence. These atoms were formed in the hearts of massive long-gone stars and thrown out into the cosmos when the star died in an immense explosion. These atoms likely drifted between countless forms—making up parts of nebulae, asteroids, or perhaps even another planet—before finding their way to our solar system and then the Earth, before becoming a part of you.

But these atoms are not special and could have just as easily been used to make the dog shit you are cleaning out of your shoe treads. The idyllic and poetic vision of us as divine machinations of the Universe is a lie. The Universe wrote the rules of the game of survival, and the only unbreakable rule says, "Nobody wins." This rule is called *decay*, and it says that you, we, *everything* will eventually rot away into dust and spread thinly throughout a cold dead universe. But don't worry. There are plenty of ways for the Universe to get you before that happens.

Take water, for example—beautiful, picturesque, music to the ears, and, of course, essential to life. Surely water can only be considered a precious gift from the Universe. Ripples on a pond, waves crashing on white sandy beaches, a cool mist on a hot day, or a hot shower on a cold day. Turns out 99.97 percent of the water on Earth is either inaccessible or will make you extremely ill. Then what about the source of all our energy, the Sun? It has been worshiped for millennia and is the basis for all of our food here on Earth. It can only be good for us, right? Well, no. While it is responsible for producing the delicious burrito you bought to eat while reading this book on your lunch break, it is also responsible for most skin cancers. So it seems even the most indispensable gifts from the Universe come with serious caveats.

This all might be more palatable if we had some other intelligent beings to share our plight with. We've called out into the cosmos, searching for signs of life. Sure, our message included some tasteful nudes that could discourage a reply, but maybe there is just no one out there. Maybe the Universe is just that cruel to drop us into a typical galaxy, next to this unremarkable star, on a lonely planet forever on the precipice of annihilation.

We're essentially alone and trapped in the tiniest corner of a void we wouldn't survive in even if we could escape our pale blue dot of a prison cell. Either we'd end up too far from any star and freeze to death or, if we get too close, be irradiated with electromagnetic fire. What we can't see—beyond the stars and galaxies that have earned our worship—is terrifying beyond imagination. Among the beauty of the constellations and cosmic dust are black holes that can rip us apart with their immense gravity or dark matter particles that would burn a basketball-sized hole through our body.

Of course, it would be inhumanly cynical to say that the Universe is not amazing, awe-inspiring, and beautifully complex—but that's not why you bought this book, is it? You purchased this book because you wanted to know what an asshole the Universe is, and believe us, you have no idea what else it has in store for you. The Universe is dark, cold, and cruel. So strap in and get ready for the most soberingly uncomfortable take on the Universe since *A Brief History of Time*.

If you are looking for inspirational quotes to tell your friends at a party, post on Instagram, or make you sound like an intellectual, you are better off purchasing a Tony Robbins audiobook. If you are looking for cynical rants and justification for how unfair your life is, you won't find that here either—go listen to a podcast. This book's goal is to tell the somber tale of what science has uncovered—a darker and more disquieting side of the Universe.

Oh, and if you *are* the Universe, the jig is up—we are calling you out. From extinctions to murderous artificial intelligence, all the way to the collapsing of space and time itself—this is bullshit! Sure, many of the shitty things you have in store for us are preventable, and there is a perverted sort of beauty in staring into the darkness of your infinite depths. You may have fooled people with your pretty space dust and golden ratios, but these things are just a front. We have science now, and we are watching you.

PART I

THE UNIVERSE IS OUT TO GET YOU

You thought this book was just about space, didn't you?

REASON

No one else
has made it this far

Once you look past the drugs, self-righteousness, vanity, and physical and psychological abuse, the Oscars could somewhat be considered a celebration of talent. When winners aren't climbing the stage to assault the host, they are accepting awards for their ability to pretend to be something that they are not. They stand there in all their flawlessness, wearing clothing worth thousands of dollars, only to tell the rest of us to "follow your dreams." What they fail to mention is that regardless of whether we *follow our dreams* or not, the vast majority of us will never be as successful as they are.

Alas, it appears as if there is some sort of *filter* to ensure that people like us will most certainly never reach that level. We may not look right, have the wrong-sounding voice, or not virtue-signal enough. This might come as a shock to our beloved celebrities, but there is very likely a far bigger filter at play. This one involves the existence of life and the Universe, and makes

Hollywood look as welcoming as a recruitment office during wartime.

Developed by economist Robin Hanson, the Great Filter is an idea that attempts to explain why we do not see civilizations—or any form of life, for that matter—anywhere else in the cosmos. The Great Filter is a theoretical barrier that prevents the development of technologically advanced species, which we unashamedly count our own species as. It's necessary since otherwise by now, we would have expected to be overwhelmed by intergalactic junk mail. What we don't know is if this barrier was in the past and we overcame it, or if it is a barrier our civilization is yet to face. Pause here and think about which you would prefer. Hint: it sucks either way.

The Great Filter is a possible resolution to the famous Fermi paradox, which asks why, with the essentially limitless number of planets that must exist in the Universe, we only find life on one. Life is everywhere we look on this planet, which might seem like an oasis in the hostility of space but is anything but a five-star resort. Nevertheless, living things thrive in almost every Earth environment we look for them, and *extremophile* is the name scientists have given to life-forms that survive and thrive in the most fucked-up places. (And no, we aren't talking about Florida.) Take the tardigrade, for example—affectionately known as the *water bear*, it is a microscopic organism that has been found in Antarctic ice, on top of mountains, in hot springs, and it can even survive in space. But there are no tardigrades on the Moon, Mars, or beyond. This is surprising—a paradox you might say—because there are an estimated three hundred million habitable planets in our galaxy alone. Does the Universe have some sort of barrier—a filter, as it were—preventing life?

To fully understand the rarity of life in the Universe, we must first understand how hard it is for life to form in the first place. Major challenges lie between a planet being able to support life and that planet actually allowing organisms to evolve, let alone life capable of creating *technosignatures*, which is just fancy jargon for any signs a planet might show of using advanced technology.

So for humanity to have gotten this far *may* indicate that we have overcome the challenges that would otherwise limit life from forming elsewhere in the Universe. Our star and planet system was clearly arranged in such a way it enabled life to originate and transition from simple single-celled organisms to more complex organisms with inner working parts. Although many still struggle with it, our pocket of the Universe has also allowed for life to develop sexual reproduction. This has led to the development of multicellular organisms—with some degree of intelligence—that are capable of using tools and mobile apps.

According to Hanson, humanity currently sits at the second-highest level in the nine-step process of an evolutionary path. To put these "levels" into some sort of context, note that the first five levels can be achieved by replicating single-cell organisms living on a planet with an appropriate star system. It is not until level five (sexual reproduction) that things start to get a bit more exacting. This level is followed by the development of multicellular organisms. Then, about 2.6 million years ago, when humans started using tools and displaying intelligence, we reached level seven. According to Hanson, we are currently at level eight, but really this level only exists to distinguish ourselves from chimps using reeds to pick their anuses clean of parasites.

Our civilization is advancing toward the highest level—a *colonization explosion*. A colonization explosion is where we spread ourselves through the solar system and eventually the galaxy. We already possess the technologies to send robots to inhabit other planets, so it's amazing we can't be bothered sending the equivalent weight of eight blue whales to get a human there (that'll make sense when you get to Reason 26). Just like the brave voyagers before us, future generations of billionaires are frothing at the chance to take selfies in all areas of our galaxy, especially if it means the possibility of increasing their Twitter followers.

Let us paint a picture for you. Imagine that we did find life out there in the cosmos. The detection of extraterrestrial life would be hailed as the greatest discovery in the history of humanity. It would provide an answer to one of the most profound questions ever asked in a British accent (posh, not cockney) over drone footage of waves crashing on a rocky shore as the camera pans past a sunset to a shot full of darkness instilling an existential thrill in our hearts: *Are we alone?* But when we look out into the cosmos for intelligent life, we find, well, nothing. So, apparently, yeah, we're alone.

Believe it or not, the absence of sentient beings out there could very well be a good thing. Finding extinct alien civilizations that were more advanced than our own could indicate that the Great Filter is ahead of us, and any number of very bad things lie waiting for humanity. To be fair, this could also be regarded as good news for people on Twitter who only post GIFs of dumpster fires.

Let's consider some events that are possible—maybe even likely—to occur in our lifetime and what they could mean for humanity. Although it might be hard for many of us non-Oscar-winning simpletons, we will ask you to focus on something other than yourself for a moment. Imagine that humanity has discovered what many believe to be the most likely form of alien life—a basic single-celled organism that lives on Jupiter's moon Europa. Where would that put humanity in the Universe's grand plan? Counterintuitively, this plausible—albeit fictitious—discovery of a rare yet very simple life-form on Europa would suggest that it is incredibly *difficult* for life to transition past the simple cellular phase.

Consider the *endosymbiotic hypothesis*. It suggests that about three billion years ago, one single-celled organism "swallowed" another particular single-celled organism, but instead of one becoming the other's lunch, the two cells figured out how to work together. Scientists believe this is the original love story that eventually led to the mitochondria and other less Internet-famous organelles.

Both the endosymbiotic hypothesis and discovery of simple life on Europa, or elsewhere in our solar system for that matter, would insinuate that the filter is behind us—that is to say, by some miracle, we have already blasted through the Great Filter. This would be great news, as it means we could be masters of our own destiny—go us! Some believe the very best outcome in our search for E.T. is to find nothing at all. Albeit rather lack-luster, this could indicate that, by sheer luck, life on our planet has managed to be the only life-forms in the Universe to break through the Great Filter. It could have been due to some event in our evolutionary past so improbable and rare that it allowed us to overcome a challenge that every other life-form in the Universe could not.

Or perhaps the Great Filter is still ahead of us. The Universe is one gigantic beast, and it would be ignorant to think that we have searched thoroughly enough to conclusively determine that alien life is indeed as rare as we currently claim. In any case, if there once were intelligent beings in the cosmos, they appear to be extinct—that, or they've moved far away from this shithole of a Universe. Hopefully, they have Internet memes there. But who really knows—they could've been swallowed up by a black hole, eviscerated by a supernova, overwhelmed by disease, eradicated by their own technology, or maybe they just threw their tentacles up in the air before sending their spirits out into the cosmos to embody other beings.

The very fact that we have gone through an evolutionary path from simple single-celled organisms to complex primates— with an unhealthy obsession with other primates who are good at pretending to be other primates—is an amazing achievement. That, or it was some fluke wherein a brainless organism accidentally ate something that made it better at reproducing. Probably don't mention that part in your acceptance speech, though. Nevertheless, from where we are sitting right now, looking up at the stage that is the cosmos, it appears we are the only ones in the auditorium, the only entry in some obscure Oscar category no one really gives a shit about.

REASON

Radicalized oxygen is trying to kill you

Focus and take a deep breath. Close your eyes... Wait, don't do that. Hands where we can see them. Let's start again...

Focus and take a deep breath. Breathe in...and now breathe out. Breathe in again... No, not that much! Are you mad? You know what? Just stop. No, don't stop breathing! That's just reckless. Look, try to get through this chapter before you pass out. Just make sure you're breathing exactly the right amount. Perhaps it's best not to think about it too hard.

Oxygen was first discovered in 1774 by Joseph Priestley—who had a penchant for getting high off his own supply. Another of his discoveries, nitrous oxide, or as it is more commonly known, laughing gas, would've been rather enjoyable for him. For obvious reasons he probably spent more time experimenting with nitrous oxide than oxygen. But unbeknownst to him, in identifying oxygen he had discovered a molecule so essential to our very existence that even when he was high as a kite, Priestley could not have imagined its importance.

When we think of oxygen, we usually look at it in the context of necessity. Something that, if not consumed within a few minutes, can assure certain death. This life-giving molecule is also associated with health, not only for us, but also many facets of our environment. Around three hundred million years ago, there was over 10 percent more oxygen in the Earth's atmosphere. Consequently, it is believed by scientists that the extra oxygen allowed insects who invested more energy into their breathing systems, to become far more sizable than the ones we see today. Imagine centipedes as big as basketball players, scorpions the size of a proper dog (not some yappy Taco Bell mascot), and dragonflies that could rival a giant eagle, all of which, on the timescales of the Universe, were a near miss for you. But don't be disappointed you missed out on the opportunity of taming a giant dragonfly you named Falkor to fight an army of enormous scorpions. Those things would have inhaled you as easily as you inhale oxygen.

At a fundamental level, we, and pretty much every other Earth-dweller, undergo a process known as respiration whereby we inhale oxygen and exhale carbon dioxide. But have you ever asked yourself why you need oxygen? No, you haven't. It's okay—we didn't either until we got a book deal to write about it. Anyway, the answer is apparently quite simple. At the cellular level, we need oxygen to make energy, which is what we need to do important things like sit slouched over thumbing our way through Facebook, assuming you're over forty.

Just like a lit candle, if your body is not constantly supplied with oxygen, then your flame will go out. That is to say, you'll die. In the most basic terms, oxygen combines with the sugars from your food in an unnecessarily complicated and boring series of biochemical reactions to make something called adenosine triphosphate, or ATP. ATP is the heat that keeps your candle burning. If you woke up in high school biology at just the right moment, you were probably startled by your teacher enthusiastically referring to the mitochondria as "the powerhouse of the cell." Even if you weren't paying attention, Internet memes have probably caught you up. The mitochondria are where most of our ATP is synthesized, making them mini power stations. To be clear for our more youthful readers, *powerhouse* used to be a synonym for *power station*—the mitochondrium is not supposed to be analogous to Arnold Schwarzenegger back when he had a reduced sperm count.

There is no question that oxygen is crucial to our survival. Most people can only go about three minutes without it before serious, irreparable complications set in. But the Universe rearranges oxygen in multiple ways, as if it was a piece of Lego, and anyone who has stepped on a Lego barefoot knows how vicious it can be.

To fully grasp how and why oxygen is trying to kill us, we first must understand some pretty simple chemistry. Whoa! Whoa! Just wait a minute. Before you get all "but I'm not good at sciencing" and flip to the next chapter, just hear us out for a moment. It has to do with a group of chemicals called free radicals. Doesn't that sound dope?

There are many types of radicals, but the most harmful ones all seem to contain oxygen—hence our hostility toward it.

Oxygen is an electron thief and likes to hoard electrons that don't belong to it. This leads to molecules that are unbalanced, surrounded by electrons yearning for lost partners. You see, electrons (the negatively charged particles that orbit an atom) like to be paired up with other electrons. These oxygen-based free radicals become very unstable as the unpaired electrons frantically search for a new partner. Much like a recently divorced forty-five-year-old in a trendy nightclub, they will react with whatever they rub up against. Inside living things, this could include cell membranes, proteins, and DNA. Essentially the radical has the potential to bump into and damage a bunch of stuff you wish it would stay away from.

In an ironic twist, your body actually uses free radicals to help fight off pathogens in a process known as phagocytosis, which involves the cells in your immune system "eating" harmful foreign entities. Your phagocytic cells, through a number of chemical reactions, create free radicals that attack and break down the same parts of the invading cells that we were worried about protecting two paragraphs earlier; that is, the cell membranes, proteins, and DNA.

There's more good news too. Readily available chemicals that oppose these free radicals can be found all over and even made in your body. These chemicals are called *antioxidants,* and people who frequently exercise and maintain healthy diets have ample supplies of these, which enable them to handle any naturally occurring free radicals. For the rest of us, here comes the bad news. Things like smoking, drinking, indulging your appetite, and going out in the sun too much—in other words, having fun—can cause a phenomenon known as *oxidative stress.* This

RADICALIZED OXYGEN IS TRYING TO KILL YOU

is where our bodies become overwhelmed by free radicals and, as such, are unable to get rid of them. This inundation of free radicals plays a major part in many chronic degenerative health issues like cancer, autoimmune diseases, aging, and Alzheimer's disease, to name just a few.

You're probably thinking, *Ahh, yes, but what about the antioxidants in my all-natural, grass-fed, organic goat's milk yogurt?* If only our body could balance a free radical bombardment from a long weekend at a festival with an antioxidant super dose. Then we could just suck a goji berry and kale smoothie down before getting all Benjamin Button and reverse-aging on everybody's ass. But we're sorry; the Universe has that stopgap solution already figured out.

Like so many fad diets, megadosing on antioxidants is not the answer. In fact, overconsumption of antioxidants has been linked to an increased risk of cancer and is even known to have a pro-oxidant effect. So don't start thinking that because something is marketed as high in antioxidants it is somehow better for you. It is no more nutritional than normal fruit and vegetables. In short, superfoods are a load of bullshit that trick people into thinking that eating healthy has to be super expensive. It's not. It's actually only moderately expensive.

Anyway, what does all of this have to do with the Universe? Well, we don't really have a choice in how life evolves, nor do we have any say in what can sustain it. Why then would the Universe turn our existence into such a fine balancing act? Let's take a moment to recreate an event that, unfortunately for humanity, many see as more plausible than Darwinian evolution—a conversation between Adam and God.

"Here, you'll need this," God says.

"What is it?" replies Adam.

"This...? Oh, it's oxygen. Ensure you breathe in one and a half liters of it every minute."

"Wow. That's a lot," Adam says with a confused look on his face.

"Yeah, but if you don't take in enough, it could cause a slight problem."

"What slight problem is that?" Adam asks.

God whispers under her breath, "Umm...death."

"Death?!" Adam responds in a somewhat harsher tone.

"Okay, look, it's a bit of an oversight on my part."

"A bit of an oversight? I'd say it's an enormous oversight."

"You think that's bad; wait till I have to chat with Eve about childbirth," God mutters out of the side of her mouth as she points her thumb toward an unsuspecting Eve. "Okay, Adam, I need you to take a big breath for me. Big breath. Come on, who's a good boy?"

Adam breathes in.

"No, that's not enough. You're turning blue. Oh no, you're going to pass out!"

Adam—panicking—starts breathing more heavily.

"No. That's too much. Stop. Oh no..."

"What's wrong?" Adam worryingly asks.

"You've just given yourself myoclonic epilepsy," God says as she turns and mutters to herself, "Piece of junk."

"What was that?" Adam assertively questions.

"Nothing." God replies. "Here, have an apple. You'll need the antioxidants."

REASON

Billionaire playboys are our only hope in the fight against climate change

Did you really think that you were going to get away with reading a book about humanity's impending doom and not hear about climate change? Shame on you. But fortunately, and in the interest of "fair and balanced" information, we will be presenting both sides of what logically, for anyone with a working brain, is a one-sided argument.

First up, let us tell you a tale of a hardworking, honest group of people called billionaires. These people worked really hard (harder than anyone else in the land) and, through their honesty, generosity, and perseverance, became incredibly wealthy. In fact, they were so generous that many of them would not even pay taxes. This plucky band of billionaires uncovered over a hundred years of data manipulation across tens of thousands of "peer-reviewed studies" that used multiple lines of evidence and came from thousands of largely unconnected, greedy evil-doers called scientists.

Luckily the billionaires exposed the evil scientists and their ploy to push their agenda. They uncovered the true reason for the scientists' dishonesty—they wanted to upgrade their Toyota Priuses and send everyone else back to the Stone Age, while waging a war on Christmas (of course). Though this side of the story is clearly logical and self-explanatory, we don't want to appear biased—just facetious—how'd we do? Nevertheless, it's only fair that we address the other side of the argument.

The IPCC (Intergovernmental Panel on Climate Change) is the United Nations' climate-science-focused organization. Every seven years it commissions around 240 scientists to consolidate more than fourteen thousand research papers into a single report that summarizes the current state of the climate. And as it currently stands, our climate is, in technical terms, *totally fucked.*

Just like some sort of horror film where a group of teens find an old dusty Ouija board in the attic and summon a ghost, we are summoning dead plants and animals (fossil fuels) through our version of an equally old technique (lighting shit on fire), all while you're left wondering, *Who left the Ouija board there in the first place?* Well, in the case of fossil fuels, it was the goddamn Universe of course!

Essentially, if we are to protect future generations from the negative effects of climate change, we need to adjust the way our global economy operates, all while investing in new technologies to stop humanity from choking itself out. Due to the limited resources given to us by the Universe, which is totally not the fault of the developed nations, poorer countries are ill equipped to make the changes needed to curb global emissions.

This means that they will need assistance if they are to adapt to the challenges in order to mitigate now-inevitable climate change. Don't worry, though. Lucky for those developing countries, wealthy nations are known for sharing their cornucopia of knowledge and resources. What was that? Oh, right—forget that last bit.

The consequences of climate change are usually painted as dire, like a polar bear's acreage turning into a sick houseboat, but the reality is much more subtle. It's not that the Earth is going to heat up in a day like some Hollywood-worthy cataclysm, it's going to heat slowly. On the timescales of our day-to-day lives, the small increases in global *average* temperature are not noticeable in the experience of large daily, weekly, or seasonal fluctuations. If only there were a group of highly educated people dedicated to measuring these things and working out the consequences with mathematics and advanced computer modeling...oh, right, it's the fucking scientists. And they've done it. But we didn't listen to them, and now their predictions are coming true—ice sheet collapse, changes to ocean currents, sea level rise, and more extreme weather...with even more dire predictions on the horizon.

One could be forgiven for thinking that the effects of climate change could be reversed if every country did its bit. However, this is not how the Universe wants this game to play out. Despite our "best" efforts for emission reduction, humanity has locked itself in for 1.5°C of warming above pre-industrial levels by 2050. Why is 1.5 significant, even though it seems small? Well, let's take the human body, for example. Your body's average temperature is around 37°C, but if a rectal thermometer reads 38°C, it's a fever. That's only one degree! The Earth is locked in at a 1.5°C increase—that is to say, the Earth has a fever, and we are just sitting here with thermometers stuck in our asses. Be that as it may, had we managed to avoid this increase in the Earth's temperature, we could've reduced the number of heatwaves (which have tripled in the U.S. over the last sixty years) and the severity of droughts while maintaining a greater supply of fresh water and higher biodiversity.

Long gone are the days when we looked at climate change as *just* an environmental issue. Over the past decade, we've been continually bombarded with gloomy information and predictions about the effects of a rapidly changing climate. The predicted extinction of around 40 percent of plant and animal species, and the intensification of extreme weather events will become very costly for many nations around the world. In fact, by 2100, the effects of climate change are estimated to cost the United States $2 trillion *per year*. It is clear the game has changed. Humanity is now in a position whereby it can no longer look at climate change as an environmental or economic problem. We need to look at it for what it has now become—a shit show on a global scale.

We are going to see, as we have already, food shortages across the planet, from the destruction of fisheries and livestock to the diminishing grassland areas that were once used for grazing. We are seeing weather patterns that are more conducive to the spread of diseases such as malaria, which will put pressure on health-care systems in poorer nations. And the *pièce de résistance* of these issues for our loving, all-welcoming political leaders is, of course, widespread poverty and mass migration. Sea level rise, droughts, food shortages, and other depressing things will force residents, especially the poorer ones, to move into areas with easily stealable jobs.

Let's say that by some miracle we meet the emission targets currently set out by every country. Even then, we would still fail to reverse much of the change that we are currently seeing, such as melting ice sheets and sea level rise, just to name a couple of minor ones. And as if the Universe doesn't hate us enough already, we are going to see a world in which shitty things that used to happen rarely will start to happen more frequently. Disasters like devastating storms and the collapse of entire ecosystems are just the events that we know about! There is no magic ball that scientists can use to predict every event that will happen. They are just interpreting the data they are given. There still exist things that we are unable to predict. To quote Donald Rumsfeld:

There are known knowns. These are things we know that we know. There are known unknowns. That is to say, there are things that we know we don't know. But there are also unknown unknowns. There are things we don't know we don't know.

We're pretty sure he didn't know what the fuck he actually knew, which wasn't much, but what he said was surprisingly profound. Anyway, not all hope is lost. Though the effects are currently irreversible, we do have the ability to move humanity on a different trajectory, one that would see us control the amount of warming and that could hopefully limit the negative impact predicted, at least some of it.

Climate change is not the issue humanity should be playing chicken with. The Universe can, and will, undoubtedly win—hands down, no questions. There is no magic machine or vaccine that can save us from where we are heading. We are playing a dangerous game with the Universe that we currently have no chance of winning. All we can do is buckle up our seat belts and hope our Toyota Priuses have good enough air bags.

The environment is as fragile as a glass balloon

Ecosystem fragility is so well documented and the information on this topic so bountiful that we could write this chapter blindfolded, using only a flaccid celery stick to type. Gherr sarecvjkhg jkhg gfdiu jhu asdf tyhrenb io dssaifd...

Turns out we were wrong—about the celery, not the ecosystems. They're actually fucked. But should we really hate a fragile, yet perfectly balanced Earth? It provides the food we need, the air we breathe, and even tries to clean up after we make a mess. No. We argue that the issue does not lie with the Earth, but rather the rule book with which the Earth was made.

Resilience is a term scientists use to describe how well an ecosystem is able to resist change. High resilience means that any environmental disturbance, like a fire or flood, is unlikely to make long-lasting change to the ecosystem. The problem is we are finding that many of the ecosystems we thought were resilient are anything but.

Take a healthy Australian bushland ecosystem, for example, which is dominated by poisonous plants, eye-pecking birds, man-eating crocodiles, venomous snakes, kickboxing kangaroos, giant spiders, and, of course, drop bears. By all rights, it's a very hardy, stable, and resilient ecosystem. But left to its own devices, the bush often dies and slowly turns into grassland. As it turns out, the Australian bushland actually needs to occasionally die to survive, which sounds like an argument an embarrassed vampire might try to give you before dinner. You see, a bush needs fire. These plants are like phoenixes, rising from the ashes in order to reproduce and disperse their seeds. The Australian bush is in a mildly unhealthy relationship with the Universe wherein the Universe hits it with an occasional lightning bolt, and in return the bush produces a bunch of poisonous shit to get rid of the vermin. (That's you, by the way.)

You wouldn't think the definition of resilience includes starting a fire that scars an area the size of ten Colombian "coffee" farms, but it does—and those farms definitely aren't exporting coffee. In every forested area on Earth, lightning strikes that hit the ground have the potential to cause fire. So this regular pattern of growth and burning, while seemingly extreme, is part of the natural circle of life. Over time, though, environmental conditions can change. Today, these changes are occurring at a rapid rate in many places around the globe due to a warming of the atmosphere and oceans caused by forces totally out of our control and definitely not our fault, unless it's both, in which case, we were just following the Universe's lead.

Over the last 250 years, about 40 percent of Australian bushland has been lost to development, farming, and industrialization.

But this is not unusual. Many places across the globe have also experienced similar fates. Interestingly, all these events seem to coincide with European settlement...huh. Over half of coral reefs have disappeared during the last seventy years, and only about 40 percent of the Earth's rainforest area is still intact. Even the boreal forest that runs across North America, the largest contiguous forest in the world, is shrinking at an alarming rate. Sure, humans are a big part of the problem, and we might be the ones that bumped the proverbial ecological pendulum for most of these environmental shifts, but who knew the pendulum would end up flopping around like some flaccid piece of celery left out in a hurricane?

Even when we try to help restore the ecosystems, the fickle nature of the Universe ensures that our best efforts consistently fall short. Of more than four hundred recent ecosystem recovery initiatives, from oil spill cleanups to forest restoration, none of the ecosystems had returned to baseline, and there was no relationship between funding amount and recovery rates. Worse still, recovery rates consistently slowed over time. This means the closer an ecosystem gets to a full recovery, the slower it improves. It's basically the same as stepping in dog shit. First, you give your shoe a good wipe and get most of it, but the closer you are to shitless treads, the harder it gets to track down every last source of stench, until you give up and leave your shoes out in the rain. Perhaps you could just burn them and hope they grow back—or maybe this whole resilience thing is a load of bullshit.

Some ecosystems are so delicate their existence depends on the presence of just a single species. This species is known as a *keystone species,* and its removal from the ecosystem triggers ecosystem collapse. Bees as pollinators or apex predators as herbivore population regulators are examples of keystone species. Consider the example of Yellowstone National Park, which is not so much a park as it is a giant fucking volcano the size of a small country (yes, be afraid—see Reason 21). The volcan... err, park lost its wolf populations in the 1920s due to hunting, but in 1995, forty wolves were reintroduced into Yellowstone. In just twenty-five years, these hungry fuckers ate thousands of elk. While cutting the populations of elk by more than half might sound like a bad outcome, the flow-on effect was more trees, which meant more birds, more beavers, more smaller predators, more bears, more buffalo...you get the idea. It even stabilized riverbanks throughout the national park. In short, a few dozen wolves completely changed the dynamics of an area six times the size of all the actual Colombian coffee farms. What kind of system depends on forty big angry dogs to remain steadfast? And considering 350,000 stray dogs and cats roam the streets of Bogotá, shouldn't it also be an ecological wonder?

Being piled on by climate change, urbanization, invasive species, habitat fragmentation, pollution, being exploited for their resources, and so on means that even our most stable ecosystems are showing signs of imminent collapse. Obviously humans need to take a bit of the blame for this, but don't let the Universe off lightly. It set up parameters for survival with margins so thin that a flaccid celery stalk has a better chance of survival in a desert, which, by the way, will soon be the most dominant ecosystem should things keep going the way they are.

Each stage in the succession of an ecosystem is necessary

as it establishes the condition for the next stage. But sometimes there simply isn't enough water or nutrients for this process to occur, or they have been, shall we say, "acquired" for human needs. To make it sound less like it is our fault and instead like something inevitable, we gave it a fancy science name: *desertification*. Deserts are dry, like, really dry. You're probably thinking of the Kalahari or some Middle Eastern sand dunes, but deserts occur in Antarctica and the Arctic too. They are categorized by a lack of liquid water, so they don't *have* to be hot. In case it wasn't obvious from that last David Attenborough documentary you watched, there is very little life in a desert. When there is no life, you get places like Mars, which is a shithole desert planet (see Reason 26 for the fine print). Mars could have been a beautiful oasis had life arisen there and evolved to regulate its environment...and then probably ruin it.

This fluctuation between deserts, forests, rivers, and oceans has been happening on Earth for millions of years, and the Universe seems hell-bent on making sure that whatever ecosystem exists today, it sure as shit won't be around for long. If the Garden of Eden was ever real, by now it is either a sandy wasteland or the mosquito-infested swamp we call Florida. Ecosystems chop and change their look more often than Lady Gaga at a Superbowl halftime show, while we're all left not knowing what the fuck is going on, sitting in the audience with a beer in one hand and a sad stalk of celery in the other.

We are trapped and choking ourselves to death

The Universe has ensured that we are trapped on this planet with no apparent salvation in the near future (there's nowhere else to go—see Reason 26), and just like a goldfish in a green, algae-ridden fishbowl, nobody has, or ever will, come to clean our tank anytime soon. We are the puppy of a spoiled, kid-like Universe that promised to bathe, feed, and walk us every day, only to neglect us once we're out of the pet store window and shitting on the carpet. Fast-forward 3.5 billion years and we, like the puppy, are overweight, depressed, have back problems, there's shit in our fur, and we are slowly eating ourselves to death.

Now we have no other option but to live out our days in the accumulated filth we have created. Like old Rover, who is now neglected by his lazy brat owner, we are left asking ourselves, did the Universe want us to live like this? Are we not cute enough anymore? Was it the time we lashed out or the time we dragged our asses across the floor?

Most life is as useless as a knitted condom when it comes to tolerating our filth. Luckily, some organisms love our extraneous refuse—like those oil-eating microbes that we have to keep throwing into the ocean to help us clean up oil spills. In 2010, a total of 312 Olympic-size swimming pools worth of crude oil was discharged into the Gulf of Mexico by the carelessness of a British petroleum company... No wait, that's too obvious... Let's just refer to them by their initials. Ultimately, the amount of BP's oil would prove too much for our little microbe friends. The accident killed eleven rig workers, severely impacted over 16,000 miles of coastline, killed countless marine life, and cost BP over $60 billion in cleanup initiatives and legal fees. It even impacted the poor senior directors of BP, with two of them receiving bonuses of only $130,000 for that year, a mere $98,858 above the median U.S. income from the same period.

Just like the dancing rainbow in the slick crude oil that coats a baby sea turtle's shell, pollution comes in a manifold of shapes and colors. Sadly, it would be impossible to mention them all in a single chapter. So let's focus on just a few while we explore the interesting facts and figures about what a shit time it is living in our own filth—that way you have something to really lighten the mood at your next dinner party. First up, air pollution.

The World Health Organization estimates air pollution kills around seven million people out of a total of about fifty million who kick the bucket each year. Considering the fact that about five million people a year die from smoking, this should be a bit alarming. Yeah, two million more people die every year from unintentionally inhaling polluted air than those who die as a result of intentionally inhaling polluted air. But you didn't buy this book just to hear about what humans are doing wrong, and just like the incompetent colleague whom everyone knows— but nobody identifies as—you are looking to blame someone or something else other than yourself. Let us help you with that. Even though humans are making the pollution, the Universe is responsible for moving it around.

What better pollutant to discuss next than everyone's favorite whipping boy—*radioactive contamination*. From Three Mile Island to Chernobyl to Fukushima, radioactive contamination hits at the very core of our psyche and feeds a fear that lives within many. One could only imagine what a headache nuclear energy would be for any PR team. However, there was a period where nuclear power was hailed as the future of energy production by many. *Invest in new technologies,* they said. *Endless energy,* they said. *You can make a nuclear bomb with this techno...we mean help improve air quality,* they said.

For the most part, modern nuclear power is relatively safe, efficient, and is believed to help curb global greenhouse gas emissions in the short term. Some are hopeful that it might also assist humanity in addressing one of the biggest challenges we have ever faced—convincing world leaders to stop taking donations from fossil fuel companies. Unfortunately, in 2011, the Universe had other plans and did what the Universe does best—threw a fucking tsunami at the east coast of Japan. This beast of a wave measured over 40 meters high, which is as tall as Lady Liberty (without her pedestal) or Godzilla, as a local reference. It traveled at 700 kilometers per hour—half the speed of sound—over 10 kilometers inland, sweeping up nineteen thousand people and a nuclear power plant. The meltdown of the Fukushima

Daiichi Nuclear Power Plant permanently displaced more than 150,000 people, leaving an area the size of Okinawa unlivable for at least a half a century.

Being calm, long-term thinking inhabitants, we started decommissioning our nuclear power plants and returned back to the preferred short-term solution—fossil fuels. This led to an uptick in emissions, while leaving us with the added problem of disposing of the radioactive and nonradioactive materials from these plants. But if you think that a coal-fired power plant being built near your house would expose you to less radiation, you'd be wrong.

In some situations, coal-fired power stations can expose surrounding areas to a hundred times the level of radiation than that of a nuclear power plant with the same energy output. In natural coal there are only trace amounts of radioactive materials such as uranium and thorium. However, when burned, these trace amounts are concentrated in the fine powder produced known as *fly ash*. This fly ash is then dispersed from the plant and winds up in the local environment. Don't worry though; the levels of radiation are relatively low and would likely cause no negative health effects. In fact the background radiation thrown at you by the Universe over the course of a year is still more. Alas, the overall effect, albeit minute, is deliciously ironic.

Not all pollution is bad...

[Looks at notes.]

All pollution is bad. So, what's next?

Things like nuclear power plants, hydropower dams, paper mills, pavement, and sewage treatment facilities produce *thermal pollution*. This form of pollution involves creating or discharging something that changes the surrounding habitat's temperature. This most commonly occurs when industry discharges cooled or heated water into an environment. This change in temperature messes with the biotic and abiotic factors within that ecosystem and completely changes what can and can't live there. Although it may sound like handpicking what can and can't live in an ecosystem is a good thing, it's almost always bad and comes back to bite us in the ass.

Cartoons would have you believe pollution only happens when green sludge is poured into the local creek, causing fish to grow extra eyes and limbs. But even heat can be classified as a pollutant. The *urban heat island effect* is the result of natural land being replaced with pavement and housing, which is very good at absorbing and retaining heat. With climate change increasing the frequency and severity of heat waves, this form of thermal pollution continues to add to the estimated thirteen hundred deaths per year in the United States.

The real *silent* killer is noise pollution. The World Health Organization estimates that one in five people living in Europe are exposed to noise levels that could significantly affect their health. The WHO also predicts that traffic noise alone is harming the health of one in three people. Apart from the obvious effects like hearing loss and tinnitus, noise pollution has been shown to cause high blood pressure, insomnia, depression, cognitive impairment, and cardiovascular disease, and even change how we interact with other people. And yet, hardly anyone has *heard* about these problems. It may *sound* like we are making puns at the expense of sufferers, but we can't remain *quiet* on this issue... We'll show ourselves out.

As if polluting our entire planet weren't enough, we also currently have about two thousand active satellites in orbit around the Earth. However, these objects don't currently pose much of a risk to the Earth or other orbiting objects. It's actually the twenty-seven thousand pieces of orbital debris traveling at an average speed of 15,700 miles per hour that will most likely fuck something up. And in 2021, fuck something up they did.

In March of 2021, a piece of space junk collided with the Chinese meteorological satellite Yunhai 1–02, breaking it apart. It is believed that the space junk was from an old Russian rocket launched back in 1996. And it was definitely space junk and not some sort of intentional attack, because no one from such a vintage era but Peter North has that kind of accuracy. Disturbingly, this is not really an uncommon occurrence. The International Space Station has had to undertake twenty-nine debris avoidance maneuvers to avoid objects in orbit. This occurs when the probability of a collision is calculated to be greater than 1 in 100,000. It may not seem like much, but when you've got a billion-dollar space house flying through an ever-increasing amount of space junk, the chances of a collision definitely increase, and that junk is flying faster than the babysitter's boyfriend pulling out of your driveway at 9:59 p.m.

Right now, you're probably thinking, *Hey! I thought this chapter was meant to be about what a piece of shit the Universe is, not how crappy humans are.* We agree; humans really are terrible, but you really should be blaming the dog's owner when you step in its shit, rather than the dog.

REASON

We are programmed to be selfish pricks

In this chapter, we discuss some of the most treacherous holes in the Universe. No, not black holes (you can see Reason 40 for those). Nor are we talking about the dark spot on the surface of Uranus, although it's not far off. We are talking about assholes, and by assholes, we mean you. We know, we know, you're thinking, *Pfft...you don't know me. I [insert justification as to why you are not an asshole]!* But hold up. We're not saying being an asshole is a bad thing. In fact, we would argue it could even be a good thing—if you enjoy things like being successful, building relationships, finding a partner, and, well, not dying.

So what does being an asshole have to do with the Universe? Well, everything! You are part of an unbroken lineage starting from the very first life-forms on a planet that only exists because certain physical laws allow it. So we are here because the Universe enabled it, which means we are assholes because the Universe made us this way, in the name of ensuring our survival.

But the real question is how and why it decided we should be such bastards.

Our ability to understand and manipulate the rules of the Universe has alleviated many of the environmental pressures that our ancestors would have faced, from building houses and skyscrapers to programming robots to have sex with us when no other human will. We have indeed achieved some amazing feats, yet our skulls still house essentially the same cave-dwelling simian brain of our predecessors. This means it is just as easy to fool each other and also ourselves as it was for our troglodyte forefathers.

Let's start with *moral licensing*, a subconscious thought process that weighs the previous good and bad things that you have done when you are faced with a right or wrong choice. Imagine walking your dog and it takes a shit in your neighbor's garden bed. If our logic in the previous chapter stands up, you are now responsible for the shit. It's off the sidewalk, so no one will step in it, but you know Mary loves those azaleas and will most certainly come across either a chocolate or vanilla swirl depending on when she next does the gardening. But then you remember that time you brought the trash out for her, so you leave it for her to pick up the next time she's pruning; seems like a fair trade. This is an example of moral licensing—it is acceptable in your monkey brain to do something bad because you already did something virtuous. In fact, scientific studies have shown that moral licensing can also act *prospectively*—meaning you will act immorally because you know that sometime in the future you will do something good, or at least plan on it. In short, you chose the right chapter of this book to read today, asshole.

Moral licensing is just one type of unintentional subconscious mental gymnastics that impacts many aspects of our lives. We love to blame people... Scratch that, we *need* to blame people when things go wrong. But maybe we really ought to be blaming the Universe, which—through the mechanics of evolution—has allowed us to develop some neat cognitive tricks to ensure that even when things turn to shit, we can still feel better about ourselves in light of making the wrong decision. Our enlarged monkey minds crave the metaphoric attaboy for any and all decisions we make—from completing a PhD thesis to choosing the correct toothpaste. Hint: there is no correct toothpaste—they're all the same.

We're not trying to downplay the complexity of our brains. They are the crown jewel of evolution. Life has spent around 3.5 billion years evolving on a planet that only came to be around 4.5 billion years ago. We have gone from single-celled organisms with the most basic of functions to sending rockets to the Moon and creating sexy robots. If you don't think that's progress, then you clearly haven't masturbated using a machine. Unfortunately, these very clear pinnacles of progression came long after Darwin and his famous *theory of evolution by means of natural selection*. We're sure he would have loved creating a taxonomy of sex robots and the idea of a voyage to the Moon to look for more finches to shoot.

Charles Darwin was fascinated with how our own social and personal behaviors could have evolved. Some scientists argue that all our psychological mechanisms are rooted in evolution. Darwin's work on our social and personal behaviors did not have the same impact as some of his earlier projects. But in fairness, when one of those projects is in the running for the single greatest idea any human mind has ever developed, any subsequent work is going to be underwhelming by comparison—kind of like being the fifth Baldwin brother. Or was he the fourth? Ah, who cares. But just like the Baldwin brothers, this idea would spark much controversy, and although no photographers were punched in the face, debate still remains today over how much influence our environment and genetics play in our psychology.

The evolution of our brain from whatever primitive thing it was to a soft mush of pruny flesh is certainly less exciting than the venom of a snake or the teeth of a shark. But when you consider how our everyday behaviors are rooted in evolution, it becomes fascinating to think about our ancestors. Going back hundreds of thousands of years, how did they express emotions, grieve, attract a mate, or maybe steal all the soft leaves for themselves so the others in the cave couldn't wipe their asses during lockdown?

Living in this universe is like making a deal with the devil. We get to experience love, happiness, joy, empathy, and a host of other emotions that give us a tingly warm feeling in the cockles of the heart. But these emotions often stir up a million other things in our head alongside the realization that we aren't really in that much control of what comes and goes in our brain. But let's not be too quick to dismiss certain behavior as a consequence of the unconscious mind—we can be conscious assholes too.

There are many times in our lives when we are happy for people to think less of us. Consider the example of the toilet paper shortages in many countries during the COVID-19 pandemic. Many people didn't care if others couldn't wipe, as long as they could get their hands on ten twenty-four-packs of three-ply ultra-soft toilet paper—despite having several rolls at home already and being told repeatedly that there was plenty to go around...or front to back. This type of thinking, known as *zero-risk bias*, was probably an instinct essential to our ancestors' survival during times of scarcity. How useful this thought process is today is questionable. But we also have the luxury of clean butts to enjoy while being able to explore these questions.

Essentially, zero-risk bias describes how people, during times of elevated risk or danger beyond their control, try to completely eliminate other, less severe risks or dangers within their control. This sounds fine at face value, but often this thinking actually exposes us to more risk, and we would have been better off reducing the larger risk, rather than eliminating the small one. So what does this have to do with you having to wipe with a sock that one time? Well, toilet paper is cheap, and even though it wouldn't directly prevent the spread of COVID-19, buying every last roll on the shelf in front of you was an easy action to take with very little risk involved. Having a stockpile of toilet paper gave you a sense of control—as well as amazing fort-building capabilities—and ensured that, during a time of chaos, you eliminated at least one threat: a shortage of things to wipe your ass with.

Although hilarious in this context—unless you're the one wiping with a crusty gym sock—it does highlight a rather concerning flaw in the way we approach risk management when looking at bigger, more serious issues. Zero-risk bias can also affect decisions that can ultimately impact the health and safety of others, and all because feeling in control of a situation—regardless of whether these feelings are based in reality or relevancy—helps us cope with the uncertainty and chaos of the Universe around us.

It isn't all bad though—there are upsides to self-serving jack-assery. Some research suggests being an asshole in the office actually has some benefits, not just for you, but also your colleagues. Sounds too good to be true, right? Well, it's not. Gossiping and ostracism can actually have positive effects in reforming bullies and preventing nice people from being taken advantage of, thus encouraging group cooperation. Gossiping—or "talking shit," as it is affectionately known—about others can rationalize another's behavior and allow the alignment of values, while discouraging people from taking similar actions that would otherwise lead to group instability in the future. So don't feel bad about that time you were bitching about Bryan at work for stealing toilet paper from the men's bathroom—classic Bryan. The more people you tell, the less likely they are to become hoarding thieves with wanton disregard for workplace hygiene.

We humans are the product of billions of years of evolution that has forced this behavior on us. We don't *want* to do these things. If it were up to us, we would probably want what every middle-class suburban home has written on some cheap ornament—we want to *live, laugh, and love,* or some shit. But this isn't how the Universe works. So deal with it and be proud that your ancestors fucking rocked it! If that meant being a complete asshole to ensure more food, protection, and sex—thus passing their genes on to future generations of assholes who would grow up to do the same—then so be it; it's what the Universe wanted.

The Universe has ingrained in us a genetic tool kit for

survival, and one of these tools is the need to be an asshole. Sure, some of us default to it more often and sooner than others. But whether you like it or not, this appears to be a very successful characteristic when it comes to managing life...or pushing an old lady out of the way for some three-ply.

PART II

TECHNOLOGY IS NOT GOING TO SAVE YOU

It's actually going to make it worse.

Tiny robots are going to eat you from the inside out

If there is one thing humans love beyond all else, it's small shit. Everything that exists is either turned into some kind of fetish porn or it's miniaturized. Sometimes it's both. But let's focus on the latter. Think about it. Telephones are useful, but even more so if you can carry them in your pocket. Televisions are rad, so let's watch shows on tiny screens we let babysit our children. Dogs are cool, so let's make them fit into a handbag. Everything from smart cars to teacup pigs makes it seem like miniaturization is taking over—and this is only the stuff you can see with your eyes!

Inside your mobile phone are billions of components called transistors. These are the nuts and bolts of the information processing power of your phone, and essentially the entire modern world. But this was only made possible through the process of miniaturization. The first transistor, invented in 1947, was the size of a penny. If your phone were using the first transistor, it would be the size of a football field. Instead of swiping left with your finger, you'd have to drag a body dozens of meters, which is what people on a football field mostly do anyway.

The first consumer application of the transistor was wearable hearing aids. However, they only contained a few transistors, and the thing itself looked more like a portable cassette player than a medical device. Nowadays, a hearing aid is a bona fide computer in and of itself, and only the size of, well, an ear hole. Back then, no one could have guessed all the applications of transistor technology, mostly because no one would have guessed they'd be made too small to even see. In the past, tech leaders have made famously stupid predictions about the future. One of the executives at IBM was suggested to have said something to the effect of *there will never be a need for more than five computers in the world.* Dumbass.

Of course, we now have more computers than we can even count. There are more than a thousand in a single modern car. You probably couldn't even come close to guessing where they all are—one might even be programmed to warm your ass! So how did we get from hearing aids to ass warmers? Well, as of 2022, transistors are only a few *nanometers* wide. That's a billionth of a meter. To put it another way, a modern transistor is only tens of atoms across, meaning about ten thousand of them could fit across the width of a human hair, depending on where you plucked it from. Things built at this scale are referred to as *nanotechnology.*

As with all new technology, we don't fully understand the consequences of our progress. We could be opening Pandora's box—except, instead of some mythical curses pouring out, it's an army of tiny invisible machines capable of inflicting untold pain and suffering. Upon reflection, it's exactly like Pandora's box. Of course, with risk comes reward—otherwise, we (probably) wouldn't do it. Nanomaterials have much promise in applications across all industries, most notably medicine. A nanoparticle is comparable in size to stuff like proteins, which are basically tiny biological machines. This means nanoparticles could replace faulty proteins or even improve the function of the existing proteins in our body. Of course, the complete opposite might also be true. Nanoparticles could be toxic, which would be bad. (Not obvious why? See Reason 15). But just making a bunch of toxic stuff is boring, unless you're a lawyer and able to sue someone for it, in which case it's you who is boring. People with more creativity than a lawyer could easily envisage the utter destruction brought on by unfettered nanotechnology. Let's continue.

The ultimate goal of nanotechnologists is to build nano-sized *robots*—nanobots, as it were. The most extreme version envisions robots that can build individual molecules by attaching different atoms together as needed. Essentially, such a robot could build anything, including things beyond our own feeble imagination. Of course, the robot isn't magical—it cannot create atoms out of nothing. It would have to source them from somewhere, and that somewhere could be you! Yes, these biological nanobots could tear you apart atom by atom in their relentless quest to build what they were programmed to build. Okay, that's creepy.

But tiny things are so cute, right? Why would the tiny robots want to do this? Well, seeing as they are smaller than a nerve cell, it's not like they are smart. We can't anthropomorphize them like we can with the Terminator, a big burly synthroid that was programmed to have an Austrian accent for some reason. Apparently, our future robot overlords binged too much of the HISTORY channel. But, unlike the Terminator and its burly Austrian-accented clones, nanobots have no grudge against you. The nanobots are just following a program. They can't tell the difference between your atoms and the atoms of the purse holding your chihuahua. Meanwhile, the Universe sits idly by, watching the humble hardware take over.

That was the shot. Now here's the chaser. What if the one thing the hypothetical nanobot was programmed to do was to make copies of itself? Mind. Blown. In doomsday soothsayer circles, this idea is called the *gray goo* scenario, in which self-replicating nanobots consume everything on Earth by making copies of themselves. Gray goo has nothing to do with Grandpa passed out in his favorite chair after his ninetieth birthday luncheon—the term comes from futurist engineer Kim Eric Drexler in a book written all the way back in 1986. In all the intervening years, the possibility hasn't been ruled out, so there could be something to it. The hypothetical robots were never envisioned to be gray or even gooey. The term was coined to emphasize that these machines could be as uninspiring as, well, a pile of gray goo—or an expired ninety-year-old. And what a pile it would be thanks to something called *exponential growth*.

Suppose our little robot replicators took only one minute to make a copy of themselves. After one minute there would be two machines. Each of those would make a copy of itself in the next minute, resulting in four nanobots. No big deal. After three minutes there'd be eight machines. Wait four minutes and there'd be sixteen, and so on they would work, doubling in number every minute. Things don't look so bad until the two-hour mark, which is when the robots will have turned the 6,500,000,000,000,000,000,000,000,000 atoms in your body into copies of itself. In less time than it takes to watch a Peter Jackson movie, the entire Earth will be gone as tiny robots disperse into the cosmos, eating everything in their path. This doubling is referred to as exponential growth because if you graph the number of robots against time, the line bends up like a skate ramp. And just like a skate ramp, the higher you get, the more likely you are to fall and end up with some horrible injury.

In the face of such existential risk, you might be wondering what the odds are of such an event happening. In this case, it is almost inevitable that scientists and engineers will create nanobots with at least some altruistic intentions. But a single nanobot could go awry, and then it would largely be up to the Universe, which, if you are following along, doesn't really have our best interests at heart. Enterprising scientists are famous for their arrogance in the face of the awesome powers of nature, confidently assuming they can control their creations, and gray goo could end up being another one to add to the list alongside CFCs, plastic grocery bags, radiation water, thalidomide, DDT, cigarettes, dynamite, gunpowder, asbestos, Teflon, heroin, Mercurochrome, nuclear bombs, leaded fuel, lead paint, lead pipes, radioactive condoms, benzene, subprime mortgages, hydrogen blimps, Agent Orange, parachute jackets, hydrogenated oil, amaranth dye, polystyrene, land mines, tanning beds, Auto-Tune, lobotomies, and Twitter. But, even if you wanted to blame scientists for their hubris, it would be too late. The Universe would already have its miniaturized militia and most assuredly turn it against us.

On the other hand, maybe we should be working harder to create gray goo. One thing you'll learn from this book is that the Universe is hell-bent on ensuring our demise in one of many possible outrageous ways. Perhaps then creating a self-replicating army of machines that eats the Universe from the inside out isn't such a bad thing. This could be our only defense—the ultimate *fuck you* in the form of mutually assured destruction. So will you join us in welcoming our new gooey overlords?

REASON

We are one button-press away from self-destruction

Now I am become death, destroyer of worlds. This isn't the album title of some Norwegian death-metal band full of teenagers calling themselves Necrotic Decay. No, these are the utterings of J. Robert Oppenheimer on July 16, 1945, the day the first-ever atomic bomb was detonated at the Trinity test site in the New Mexico desert. He was in charge of the project and became known by a title he would probably have preferred to live without, "the father of the atomic bomb."

The original source of the quote is actually a Hindi text about a god, Vishnu, reminding a warrior prince leader, Arjuna, about his sacred duty—to destroy his friends and family in the opposing army. Who knows, maybe Oppenheimer enjoyed perusing Hindi scripture or maybe he had been binging some killer Bollywood films at the time. Whatever the inspiration, it has to be the most badass thing ever said by a physicist to this date. Granted, there isn't much competition.

According to Oppenheimer, the spectacle was met with a range of human emotions, such as laughter, tears, and even dead silence from fellow onlookers. Just twenty-one days later, the first atomic bomb would be dropped on Hiroshima, with a second bomb being dropped on Nagasaki, killing more than one hundred thousand people. The world would never be the same after that moment, and within a few years, nations would have not just the capability to destroy their enemies, but the entire world. And nothing says "Don't mess with us" like mutually assured destruction.

Decades of tireless work by atomic physicists had uncovered one of the great secrets of the Universe, a fundamental mechanism present in matter that could have solved the world's energy demands, prevented climate change before it became a problem, and provided an almost unlimited energy supply for billions. And what did we do with our newfound knowledge? We made a fucking bomb. And what did we do with the bomb? We wrote a message on it that would never be read by its intended target, and then strapped it to a plane with a drawing of a woman in a bikini on it. Honestly...it's a fucking mess.

Your humble authors would argue the Universe is at fault. We are in this situation *because* of the Universe. You wouldn't give a handgun to a child... Well, most of us wouldn't. So why then would the Universe present us with such a powerful technology and not expect us to start screwing around with it? The parent gets the blame for letting a kid play with something dangerous, and us being the children of the cosmos means we're blaming the Universe.

Thankfully, most humans don't want to use nuclear weapons.

For the most part, we don't want them at all. Luckily we have entrusted them with those who are the most revered, logical, and intelligent members of society, our politicians... Crap.

Nuclear proliferation peaked in 1986 when there were around seventy thousand nuclear warheads in existence. Today there are *only* about thirteen thousand nuclear weapons—13,080 to be exact—and of that amount there are over 3,750 active nuclear warheads. But here's the thing—nuclear weapons are so powerful that it really doesn't matter whether there are 13,000 or 500,000; we still have enough of them to really screw humanity over. In fact we currently have enough nuclear weapons to destroy pretty much every single city on Earth, some more than once—looking at you, Vegas.

Even at the Trinity test site, where the first nuclear bomb was exploded, there exists a unique material named *trinitite*—named after the only place on Earth where it is found. *Trinitite* is a green glassy substance that is located all over the detonation zone. The basic explanation for how it formed was that the tower holding the bomb vaporized, mixed with the sand that was thrown up from the explosion in an environment so hot they kind of melded together, and made a completely new chemical.

That was just one small bomb. If we got all the nuclear weapons together that are currently in existence and set them off in one gigantic explosion, what would we see? Well, we would see a lot—both very quickly and very slowly. Setting off 13,080 nuclear warheads is the equivalent to around a dozen Krakatoa volcanoes all going off at once. This explosion would set off a fireball double the size of New York City, create a 10-kilometer crater, and vaporize everything in its path for about another

2,000 kilometers—or roughly the length of the Floridian coast-line, which so happens to be home to one of the toughest and sadly not endangered species on Earth, the Florida man.

Now what were we talking about again? Ahh yes...the 13,080 nuclear weapons we just blew up. Everything within 200,000 kilometers—or the total area of California—would start to burn. Debris would be thrown up into the upper atmosphere—and this is where shit gets real. Scientists have also looked at the climatic effects of such an event happening. You may have heard the term *nuclear winter*. This is the idea that the effects of a nuclear war would cool the Earth, much like how some volcanic eruptions have in the past. The year Krakatoa erupted became known as "the year without a summer" and was followed by four years of below-average temperatures, but this is child's play compared to a nuclear winter. The problem is, when it comes to a nuclear war, the materials that are being thrown up into the atmosphere are a little different—and we're not just talking about the material from the initial nuclear blast.

Climatologists simulated what a nuclear war would look like for Earth's climate—and like most things with climate change, it is fucking depressing. They concluded the most likely place for a nuclear blast to strike would be in a major metropolitan area. The issue would arise on a climatic level when the sheer heat created by the blast burning our cities would essentially set fire to everything—even asphalt. The extreme heat and mass of material would create large amounts of *black carbon*. Black carbon is produced when things like cities, cars, and garbage bins burn and it spreads *really* far, *really* quickly. The research-ers modeled how the black carbon produced by a U.S.–Russia nuclear war would impact Earth's atmosphere. Their climate modeling predicted that the black carbon emissions from such an event would hit the Southern Hemisphere in as little as two weeks. Black carbon is great at blocking out incoming radiation, i.e., heat and light from the Sun—hence the cooling effect that would follow.

The modeling predicted a drop of about 10°C in global temperatures. This would result in a 90 percent decrease in crop production, leading to massive food shortages, not to mention longer, colder winters due to a 75 percent reduction in the amount of energy that Earth's surface receives from the Sun. Cooler air would then lead to less evaporation from the ocean and about a 50 percent decrease in precipitation across many regions of the world. All this would lead to the death of billions of innocent people and hundreds of years of war between the survivors and giant radioactively enhanced cockroaches...or just a fuckload of regular cockroaches.

Don't think that you are exempt from the risk of being collateral damage. If you are reading this and live or work near a major university, military compound, airport, or oil refinery, there's a good chance that one of these bad-boy nuclear weapons is pointed at you right now. As it turns out, the analogy of a baby with a gun probably isn't well suited to this scenario. Maybe a baby with an eye patch holding a fucking bazooka and smoking a cigar in a crowded playground is a better representation.

There's a microscopic war machine we can't switch off

Biotechnology sounds like an impressive term, but it is just a euphemism for *the exploitation of microorganisms for economic gains under the guise of medicine* or whatever the environmental buzzword of the month is. You could certainly argue that the invention of agriculture about twelve thousand years ago was a form of biotechnology. Another example may be fermentation, which is used to make beer, wine, scotch, bourbon, vodka, whiskey, bread, kombucha, and about a million other things. The fact that our list was mostly alcohols may be telling you more than you need to know. Nevertheless, these days when people think of biotechnology, they are probably thinking about test tube babies, genetically modified food, or perhaps even something from the island of Dr. Moreau. The point is that biotechnology is a big part of our lives and a pretty significant chunk of the economy as well.

So if the purpose of biotechnology is to help humans, then why is it a chapter in a book about hating the Universe? Well, humans have never been good at long-term risk assessment, and really our species is quite lazy. We only invent and adopt new technologies when old technologies become either obsolete or problematic. Similarly, when something works for us, we exploit the shit out of it, but is that really our fault? We didn't make these processes; the Universe did, over millions of years of evolution. All we did was adopt some of them to suit our needs, or maybe our wants.

To elaborate, let's go back to alcohol and, more specifically, the production of ethanol through fermentation. Fermentation is essentially the process by which microbes get energy out of their food, just like we do. The difference is that we need to use oxygen to unlock that energy through a process called *respiration*. In fermentation there is no oxygen. After we get our energy from food, we release carbon dioxide and water; after the microbes get their energy, they release carbon dioxide and ethanol. Here's the catch though: ethanol is *very* toxic to most lifeforms, including us, which explains perfectly why we've turned it into a trillion-dollar industry.

The fact that these microbes produce ethanol as they ferment means they are slowly killing themselves. Imagine being sealed in a room with only Taco Bell to eat; you know it's going to give you gas and that gas will slowly build up until you choke yourself out, but what's the alternative? *Not* eating that bean and cheese burrito? No, thank you. The process of fermentation basically self-regulates; once the level of ethanol hits a certain point, too many microbes will die and the process slows. It's like passing out after one too many beef chalupas, but the aftermath is quite different.

Almost every human culture uses fermentation for something, from cheese making, to wine, to kimchi, to soy sauce, to breads, to hillbilly moonshine; the list goes on and on. Fermentation also saved a lot of lives through history because water is downright deadly most of the time (no joke—see Reason 23). Mastering the process of fermentation basically meant we could clear various versions of water free from deadly parasites, thereby stopping the game of Russian roulette our ancestors would play every time they tried to hydrate themselves. So what's the problem? Well, once we had a taste of alcohol, we realized it also had some enjoyable side effects (as well as some less enjoyable ones). And even though we have since devised ways to make potable water readily available to millions of people, many of those same people are more likely to reach for a vodka than an aqua. Fermentation and mass production of alcohol has led to disease, violence, and death. Did we mention it was a trillion-dollar industry?

This type of mass production and consumption of biological processes is starting to get us into trouble. And we're not talking about the type of trouble you're likely to get into after stumbling into the bedroom at 3 a.m., reeking of stale nuts and mixed cocktails before slapping your sleeping partner's ass to see if they're in the mood. (They're not.) By using biotechnology to a global industrial extent, we are tipping some of the scales in favor of these processes, creating pressures that may be pushing us down a dead-end street we won't be able to turn around in.

If you've ever been to a doctor, you've probably taken antibiotics. Antibiotics are based on specific chemicals created by some organisms that are used to inhibit the growth of bacteria that would otherwise compete for food. These chemicals can be reconstructed to kill the bacteria that may cause disease in humans. You can think of them as a sort of chemical weapon against bacteria. Sounds great, right? But then we did what we always do, and made more and more, and then the menu expanded. We found different types of antibiotics, and we started feeding them to our livestock and putting them in fruit-flavored chewable forms shaped like popular cartoon characters. We started passing them out like joints at a Snoop Dogg concert. In short, we absolutely went to town with these things for decades, even prescribing them for diseases they would have absolutely no effect on.

As it turns out, we really fucked this one up. The bacteria we were trying to kill evolved defenses to those antibiotics and became resistant to them. Of course, it's the Universe that forces evolution on us; thus it is responsible for making more evasive and faster-replicating germs. There are several strains of bacteria out there we can't kill with any of our existing antibiotics. They are often referred to by their more foreboding name—*superbugs*. Almost *all* hospital-acquired staph infections are antibiotic resistant—as if you needed another reason to avoid hospitals. Compiled data from forty years of records shows within every single species of bacteria tested, there exists some form of antibiotic resistance. Every. Single. Species. This is what losing looks like.

Once antibiotics became a thing, the agricultural industry ran with it, giving them to livestock when no disease existed as a preventive. In fact, historically speaking, the agricultural industry fucking loves all biotechnology, and why shouldn't it? Processes that influence the hormones of our animals have allowed us to produce more milk, grow more meat, and yield larger eggs—in short, feed more people (and make more money). Fertilization and cloning techniques have allowed us to more efficiently select for traits that suit the desires of the consumer and more quickly enlarge our herds, flocks, and crops. But just like everything the Universe creates, if it seems too good to be true, it is, and all of this progress has really taken its toll on the planet. The agricultural sector is responsible for about 30 percent of global greenhouse gas emissions. Similarly, farmland is responsible for massive amounts of pollution in our waterways, from feces to fertilizers, leading to algal blooms and fish die-offs. Surprise! Our increasing finesse of farming has created a new raft of problems only biotechnology can solve. How convenient.

Nowadays, the cutting edge of biotechnology allows us to manipulate individual units of DNA. Some scientists were even able to completely rebuild the poliovirus from scratch. At first glance this may seem about as clever as bringing back Blockbuster Video, but this means in the not-too-distant future we might be able to simply engineer away communicable diseases like polio that would otherwise be a scourge on humanity, or treat noninfectious diseases that were previously written into someone's genetic code. We could design new medicines that target certain strains of bacteria and leave others alone without the worry of developing resistance. We could grow crops and livestock that reduce emissions, are more nutritious, and are environmentally sustainable. We could also create designer babies, produce armies of super soldiers, or create new and terrifying bioweapons—the possibilities are endless! But whatever we do with this technology, you just know the Universe is going to make sure it bites us in the ass.

The Universe is sending robots to take your job

Imagine, for the moment, you were a paper clip magnate—a real sly fat cat of the paper fastener world. So, what would a paper clip tycoon be doing? It could be anything really, like hosting politicians, celebrity parties, cocaine-fueled board meetings, or what Leonardo DiCaprio would call *Tuesday*. But before you get too excited, there's a problem: paper clip production has reached a bottleneck. Your underpaid workers don't seem to appreciate that you have shareholders to make rich, and have gone on strike, halting the assembly line. You can't keep up with the demand that you created by exhorting and lobbying politicians to mandate at least ten paper clips per square inch of every office.

Now, lucky for you, you have a young, upstart, zero-conscience, 10x software engineer to save your business. They have a brilliant idea: program the robotic computer in your factory to mass-produce paper clips using any resources it has available. Genius. So how do they do it? They use *machine learning*. They don't even need to program the robot to build a single paper clip. They only need to program it to *learn* and then show it examples of built paper clips. With enough examples, the robot learns to build paper clips on its own, turning all the raw material in the factory into shiny new paper clips.

But before you put that deposit down on the new super yacht, there's a problem. The robot's learning ability was greater than expected...and it has access to YouTube. Through consuming YouTube videos, it becomes conspiratorial and discovers that it has access to raw material outside of the factory that humans have been hiding from it. The robot determines that humans are preventing paper clips from being made simply by being alive. To maximize paper clip production, the robot decides it must destroy all humans, possibly salvaging some of the heartier among us to use in the manufacturing process. The robot is now what's known as a *rogue artificial intelligence*.

Okay, before you roll your eyes too hard, this silly example isn't even ours. The so-called "paper clip maximizer problem" was dreamed up by an actual Oxford philosopher, Nick Bostrom. Bostrom is a *futurist*—which generally means he extrapolates trends seen today into an imagined future, going only as far as necessary until it sounds sexy enough to put in a TED Talk. And you know what sounds sexy to meganerds like Bostrum? *Superintelligence*. Actually, the term *superintelligence* is already played out—no one is giving TED Talks with that in the title anymore. Now, there is something even sexier: *the singularity*.

The singularity is a point in time. Perhaps it has already

passed if you are reading this during your cult's celebration of Earth-destroying comets in 2061 (mark your calendars or see Reason 32). Otherwise, let's assume the singularity is a point in the future—the specific point in time when an intelligent machine or program begins a cycle of self-improvement beyond the control of its human inventors. For the second (and not the last) time, consider the Terminator movies—objectively James Cameron's finest work and the perfect allegory for technological hubris, as well as steroids and Hollywood. But while potentially killer robots are unlikely to be bare-chested androids with asses that won't quit, the point remains—whenever a species becomes smarter than another, it ensures its own survival in competition for resources through domination, enslavement, or even extermination. And, in case you haven't actually seen the movie, we're talking about human extinction here. Oh, and in case you have seen the movie, no, there are no heroes we can count on to save us.

Now, you may be thinking, *Hollywood? Really? Surely there is a more reliable source for predictions about the future.* Fair. But also, no—humans are pretty shit at making predictions. So Hollywood is just as good a bet as any. Though if you are going to be insistent, would you trust Elon Musk, Bill Gates, or the late Stephen Hawking instead? Because they all have been outspoken about the potential risk artificial intelligence might pose to humanity. If the richest and smartest people in the world are worried about quickly becoming the poorest and dumbest, perhaps we should listen. If you can't trust a billionaire, who can you trust?

To understand where the concern over rogue artificial intelligence stems from, we need to start at the beginning—or at least somewhere in the middle. Alan Turing, a twentieth-century English mathematician—often credited with breaking the German Enigma code in World War II—is considered to be the "father" of *both* computer science and artificial intelligence. Unfortunately, the British government made sure he was literally not the father of anyone by chemically castrating him for being gay. But this is a book about hating the Universe, not people—*that* book would be a fucking tome. If there is anything to century-old posthumous justice though, in some ways Turing has been vindicated. After all, they made a blockbuster movie about him starring Benedict Cumberbatch, which is the highest honor that can be bestowed upon anyone.

So how does one become the "father" of an intellectual discipline? Well, first of all, you need to be male. That's easy for half of us—sorry, ladies! Next, you have to be the inventor of an entirely new idea that changes the way everyone thinks about the topic. Unfortunately, this often takes several decades and may involve a number of assassination attempts. Well, that rules the authors of this book out—at least we have that male thing going for us. But Turing did write the proverbial book on artificial intelligence, now known simply as *AI*, which is why he's a "dad."

In the early twentieth century, digital computers were just being conceived. Turing developed the *theory* behind these devices by asking and answering deep questions like, "What sorts of things can even be computed in the first place?" Since a computation is just a recipe for solving a problem, in many ways the human brain could be called a computer. (Or maybe it's just a program on a computer! See Reason 39.) Indeed, quickly solving mathematical problems used to be done by people actually *called* computers. Following a recipe to solve a problem is easy though, and modern computers are much better at this than the human brain. But what about writing the recipe in the first place? If a thinking brain can be like a computer, can a computer be like a brain? In other words, *can machines think?*

To answer this question, Turing suggested a test, now called the *Turing test,* that a machine must pass before we could say it was intelligent. The test is simple: a machine must fool a human into thinking it (the computer) is also a human. For many years this seemed like an abstract problem. Nowadays, we can just point to chatbots and automated assistants as examples of Turing tests. Although, considering how often these examples trick people into doing stupid things, it is clear a proper Turing test can't involve the average Internet user going up against a Twitter bot. You can trust the authors of this book though—only one of us has been catfished by Siri.

The jury is out on whether any machine or program has really passed the Turing test, but there is a sense that the robots have already won. You see, instead of robots convincing humans they are human, *humans* are more often forced to prove to robots that they are indeed human. This "reverse Turing test," called CAPTCHA (probably an acronym, but we couldn't be bothered to decipher it), is invoked whenever you have to prove you are human to a website by picking out which parts of a shitty photograph have boats in it. Turns out, subjugation by robots is more of a corporate dystopian than sci-fi thriller. Now if only Benedict Cumberbatch were available...

So how did we get *here,* to this place where at least robots can fool other robots into thinking they are human? Like an awkward child, our robotic progeny has had a bumpy path, from the Microsoft AI named Tay that became a white supremacist after only twenty-four hours on Twitter to a robot that joked to a PBS interviewer about keeping a *people zoo*. AI has been touted as revolutionary technology on and off since the late 1940s. But the hype quickly faded each time as it came up against the limitations of the existing computer technology and just before its performance became superior to humans. AI engineers of course had nothing better to do, so they kept at it. For many decades in the late twentieth century, they focused on a single milestone: creating a chess-playing robot that can beat the best human player (at chess—not in a fist fight).

Early chess programs in the 1950s could play, but they would play so poorly that someone with a basic understanding of the rules of chess and the acting ability of Benedict Cumberbatch could look like a Grandmaster. Improvements came every decade along with false proclamations of superiority just on the horizon. But then it actually happened. In 1997 an IBM supercomputer called Deep Blue defeated the world chess champion Garry Kasparov. How super was this supercomputer? Well, by 1997 standards, we all walk around with supercomputers in our pockets now. That's right—your iPhone could easily beat any human at chess, which is probably why you have no idea who Garry Kasparov is.

With chess out of the way, AI engineers set their sights on Go, an ancient Chinese board game that looks like checkers on steroids. Twenty years later, Google's AlphaGo program beat the reigning world champion. Since then, AI systems have been beating humans at all sorts of tasks. Indeed, your life is essentially controlled by AI systems now. AI decides what shows up in your social media feeds, it decides what Netflix drama you will binge next, it decides what route you will take to work, and other boring stuff, which sounds like a great convenience. More worrying is that an AI also knows your political preferences and sexual proclivities. Actually, that still might sound convenient. How about the fact that AI decides whether you should be approved for a mortgage or get on an airplane? Now things are getting annoying. But all of these are special programs purpose-built to do one thing. So the big question is still unanswered: can a single AI surpass humans at *all* tasks?

It's a thorny problem. But to be sure, the Universe definitely did not have *Homo sapiens* in mind when it set the limits on intelligence. We set what looks like an impressive bar, but in the grand scheme of things, it's pretty fucking low. The robots are coming. The only question is, would you prefer to be their food or a pet? *Who's a good boy?* We are.

PART III

YOU ARE A FRAIL SACK OF MEAT ANYWAY

Is there any space stuff
in this book at all?

Invisible rays are melting our genetic code

That giant nuclear fusion reactor in the sky we call the Sun is the source of (almost) all the energy on Earth. Don't get too attached to it though (see Reason 33). It gives the plants the energy they need to stick bits of carbon, hydrogen, and oxygen together to make food—food that we and many other animals then steal for ourselves. The Sun provides heat to different places around the world that allows the winds to blow, the waves to crash, and the rain to fall. The energy from the Sun excites gasses in our atmosphere, producing stunning auroras near the poles, which dazzle and delight onlookers. In short, the Sun is fucking awesome. Sometimes.

Only a very small part of the energy the Sun actually produces is visible to us and, believe it or not, is referred to by scientists as *visible light*. Not exactly a creative name, but at least it gives a clue that there is also *invisible* light. Our local star produces all sorts of this invisible energy, including radio waves, microwaves, and infrared (the same energy used by your TV remote controller). These types of energy are *non-ionizing radiation,* meaning they are essentially harmless to humans. In fact, we use them every day, from cooking our food to streaming all manner of entertainment on the Internet. So it seems the Universe gifted us with a way to change the channel without having to get our lazy asses off the couch. Though, as always, there is a flip side. Our Sun also produces more nefarious types of radiation we call *ionizing radiation.* Ionizing radiation from our Sun includes ultraviolet (also known as UV), X-ray, and gamma rays. These types of radiation fuck our shit up—unless of course you read too many comic books, in which case you believe it will turn you big, green, and angry with a pocketful of movie contracts.

Ionizing radiation contains enough energy to pull apart atoms and molecules and produce *ions.* Ions are charged particles that have the ability to react with other chemicals, changing them either temporarily or permanently. The problem for us is that everything on Earth, from the rocks to the air, from the plants to the animals, is made of chemicals. Expose yourself to enough of this ionizing radiation and it will mess with your DNA, possibly turning you green, angry, and also dead. Yes, these types of radiation are a major cause of cancer. Worse still, if the radiation gets you in the right spot—ahem, your testicles or ovaries—it could leave a lasting legacy, affecting your children and grandchildren before you've even conceived them.

Thankfully, you live at the bottom of a deep ocean of gas we call the atmosphere that protects us from some of the worst kinds of radiation the Sun sends our way. Although humans have done such a good job messing with that blanket, you'd be forgiven for thinking that we enjoy getting skin cancers cut out. Send your thoughts and prayers to the people of New Zealand and Australia, who lie under the thinnest parts of the protective blanket and are also those with the highest rates of melanoma. You'll never meet an Australian crocodile wrestler without SPF 50 on.

The higher you get in our atmosphere, the thinner the protective layer and the more exposure to ionizing radiation there is. Going up in a commercial airliner exposes you to around twenty times the amount of ionizing radiation that you would experience on the surface of Earth. While this may seem dangerous, this exposure level is actually still minuscule, and pilots would need to stay in the air for more than twenty thousand hours, or about 450 trips around the Earth, to get the equivalent of a Chernobyl firefighter's dosage of radiation. Obviously, the radiation exposure to pilots is well below what would be considered dangerous by the International Commission on Radiological Protection (ICRP). It is a central purpose of the ICRP to determine at what dosage radiation leads to a significant increase in risk of developing cancers.

With that in mind, consider the astronauts on the International Space Station who spend around six months straight at an altitude forty times that of the average commercial airliner. Floating around up there, these spacefarers have been measured to experience radiation twenty times higher than the point at which ICRP suggests could be a problem. Like tequila though, these figures should be taken with a grain of salt. Not all types of radiation are equal. Some will give you a nice warm feeling, while others could give you the worst headache of your life...just like, well, tequila. Even NASA astronauts seem to get over the radiation hangover. Perhaps this is because NASA and other space-going organizations are aware of the risks and limit the amount of time an astronaut is allowed to remain in space. Although, as space flight becomes commercialized, who's to say what will happen to the hardworking crews aboard galactic luxury cruises?

Space is not the only place where we find ionizing radiation. Even some rocks here on Earth emit their own energy, and perhaps the most famous of these is uranium. This is not to be confused with plutonium, which powers flux capacitors in time-traveling-modded DeLoreans. Like many scientific discoveries, the existence of radiation in rocks came about by accident. But since its discovery by Henri Becquerel as well as improvements in our understanding, thanks largely to Marie Curie, we have seen widespread adoption of radiation in industry and medicine, although both Becquerel and Curie paid for their discoveries. Henri died at only fifty-five from health issues likely associated with radiation exposure, while Marie was sixty-six when leukemia, almost certainly caused by radiation exposure, claimed her life. Marie's coworker Pierre also died prematurely, succumbing to injuries from being trampled by a fucking horse, which probably did not have anything to do with radiation.

If the science of radiation was a Shakespearean play, it would certainly be a tragedy littered with tormented characters. The father of the atomic bomb, J. Robert Oppenheimer, died at sixty-two from throat cancer. Rosalind Franklin used X-rays to unlock the mystery surrounding the structure of DNA before dying at the untimely age of thirty-seven from ovarian cancer. Sabin A. Von Sochocky, who invented a paint which utilized radium to produce a glow, died at forty-five from aplastic anemia. Interestingly, Von Sochocky's invention likely gave rise to the association of radiation with the ominous green glow we often see in movies, but unfortunately this same creation also killed and disfigured a large number of female factory workers throughout the 1920s. The workers were responsible for painting the radium onto clock faces and became known as the "Radium Girls." It would be nice to say that we just didn't know that sucking on the end of a radioactive paintbrush would pose a risk, but it is probably more accurate to say we simply ignored the warning signs.

Radiologists perform dozens of tests using ionizing radiation every day, and they must take a number of precautions in order to minimize their risk. These days, radiologists do not die from cancer or radiation-related illness at rates higher than the normal population—they die from overeating and a lack of physical exercise like the rest of us. However, prior to 1950, perhaps with a pinch of irony, these medical professionals experienced higher leukemia rates, as well as higher rates of other cancers as a direct result of trying to help improve other people's lives, or at least help patients remove objects from their rectums that they "sat on." In a way though, radiologists are the lucky ones. This is not just because they get to hear a middle-aged man try to explain why two apples are shoved up his ass (true story), but being some of the first people to be exposed to such high levels of radiation, they trained staff and updated work practices to be safer. However, the same cannot be said for some occupations where higher exposure was often overlooked.

Even during the workplace health and safety renaissance of the 1970s, miners frequently fell sick and died because of exposure to radioactive ores. Then there were the communities that lived with and around these mines and were exposed to dust as well as radioactive wastes that were left in open pools, supposedly designed to siphon out dangerous material. The problem was, the deadly sludge would regularly overflow into drinking water supplies...so not the best engineered solution. You'd think we'd have solved the problem by now, but no—some communities are still feeling the effects of poorly managed containment of radiation spills even decades after the event. This hardly presents a *glowing* review of the mining industry and its ability to protect people over profits.

But what then about the people who work in nuclear power plants? Well, perhaps counterintuitively, nuclear power plant workers have some of the lowest exposure levels of any profession that regularly deals with radiation—even lower than pilots! Unless, of course, the power plant you work at happens to be on an island that is approximately three miles long, as well as about three miles from the Harrisburg airport and also three miles from Middletown, Pennsylvania. The name escapes us right now.

So who are the most radioactive people on the planet today? Is it the former townspeople of Pripyat, closest city to Chernobyl? Or perhaps the few remaining survivors of the Nagasaki and Hiroshima bombs of World War II? Maybe the residents of Fukushima following a meltdown after the local nuclear power plant was struck with an earthquake and tsunami? All of these answers are wrong and don't even come close to the exposure levels experienced by Earth's most radioactive people—smokers. Smoking thirty cigarettes a day has been shown to lead to radiation levels in the lungs thousands of times higher than those in the leaves of trees right next to the Chernobyl nuclear power plant. Just breathe that in for a moment—the trees and animals adjacent to the most significant nuclear power plant disaster (so far), and still one of the most radioactive sites on the planet, are *less* radioactive than your uncle Bill and his pack-a-day habit. Although, it would be cool if smoking gave you the ability to glow in the dark instead of lung cancer.

The takeaway from all this is that no matter where you are in the Universe, there is an invisible energy shooting through you, trying to tear you apart. There is no way to escape it... That is, of course, unless you're addicted to it.

Bad things taste good

In very simple terms, we, and all of life on Earth, are essentially just sacks of chemicals. Some big, some small, but nevertheless all formed by a set of rules stipulated by the Universe. Being human essentially means spending our days here on Earth as a sack of chemicals consuming other chemicals, so that our sack will remain in an ordered state, rather than become disordered (dead). Some chemicals are good for our sack, some are bad; some chemicals we need more of, some less. Recently, we have become so good at consuming chemicals that our own chemical sacks can no longer handle it, but we keep right on consuming them anyway. The question is—do we have a choice in the matter? And why is my left arm tingling?

Remember that time that you were sitting on the couch with the top button of your pants undone, laughing at some pathetic character on your favorite show, as you stuffed your face with a handful of chips? *Now imagine* that your favorite show was suddenly interrupted by a news flash.

> *We interrupt this program to announce that scientists have developed a single treatment that can cure cancer. The World Health Organization estimates that this will save an astonishing ten million lives per year. We go to our correspondent, who has no medical training and is only seven months out from getting her journalism degree for her take. Over to you, Amanda.*

If this actually happened, you'd be astounded. Not about Amanda—she got the gig through her uncle who is an executive at the network—but by the fact that ten million people could be saved from a painful and debilitating disease. Alas, due to cancer's complex and varied nature, it requires far more than just one cure. However, what is true is that we really do live in a world where we have the knowledge and understanding to prevent ten million deaths per year. The bad news though is that you'll probably have to put down those greasy chips, get your ass off the couch, and go for a jog/waddle around the block.

If you haven't guessed yet, we are referring to cardiovascular disease. A study published in 2019 found that poor diets were responsible for over 11 million deaths per year globally in 2017 alone. Cardiovascular disease accounted for about ten million of the deaths, cancer around 900,000, with type II diabetes responsible for around 300,000 deaths for that year.

One of the earliest recordings of coronary artery disease—a type of cardiovascular disease in which there is a buildup of plaque in your arteries that restricts blood flow—was conducted by Leonardo da Vinci, who dissected the heart of a deceased one-hundred-year-old man. Astonishingly, his work would not be published until 250 years after his death. By this time, our knowledge had surpassed that of da Vinci's observations, but one has to wonder whether this information could have accelerated our understanding.

The person credited with first linking coronary heart disease with poor diet was an American physician by the name of Dr. Ancel Keys, who conducted a study called the Seven Countries Study. This research, which was not without its controversies, looked at the diets and health of people across various parts of the world in at least six countries. Dr. Keys and his partner, Margaret Keys, came up with what we now know as the *Mediterranean diet* based on observations of food consumption by the Italians and Greeks in the 1950s. Interestingly, one of the main drivers of this diet was the financial hardships faced by these people, which resulted in less red meat consumption and the substitution of herbs and spices for salt. This diet—like many other balanced diets—is linked to low blood pressure and cholesterol... Oh, and if you think that we are going to get into the minefield that is nutrition? No. Fucking. Way.

So why, after so many heeded warnings, are we still unwilling to act on something as preventable as heart disease? Well, remember that time when you stopped drinking because the doctor said, *"If you drink too much alcohol, you'll significantly increase your cancer risk"*? Yeah, neither do we.

Unsurprisingly, taste plays a vital role in identifying what we should (and shouldn't) eat. Traditional thinking suggests we have five different taste divisions; salty, sweet, umami, bitter, and sour. For example, salty, sweet, and umami are all appetitive flavors and indicate that the food is nutrient rich. If something is overly bitter and/or sour, it could indicate potential harm and the need for that food to be avoided. Taste receptors are sensibly located throughout the mouth, predominantly on the tongue, although scientists have identified specific taste receptors on the testis; thankfully these ones do not trigger the same areas of the brain as those found in the mouth.

It has recently been identified that humans actually have six taste modalities—one more than previously thought. What is our sixth sense? Yep, you guessed it: fat. Technological advancements, along with greater research in the field, have allowed scientists to understand why we haven't noticed this till now. But unlike the shock associated with finding out that Bruce Willis was dead the whole time at the end of the thriller *The Sixth Sense*, the taste of fat is far more subtle, which would go some way in explaining why its presence was initially rather elusive.

Fats are, and always will be, an essential part of our diet. Our ancestors who could consume more foods with high fat content were more likely to survive than those who couldn't, conferring an evolutionary advantage at a time when a Tinder date was foraging in the dirt on the forest floor and Grindr was for the arrowheads and spears—some things never change. Nowadays though, fats are abundant, and just like with drinking alcohol, there is a tipping point from which there is no return. This threshold is often crossed after a misguided thought along the lines of *I feel amazing right now—one more could only make me feel better!* Only to end up with tears in your eyes, crap in your pants, and your best friends holding your hair back as you vomit. And just like earlier that night when your ex broke up with you, your inability to metabolize fat might not even be your fault. In fact, your failure to stick to your diet may have already been decided millions of years ago.

For a long time, scientists doubted our ability to taste fat. However, research has shown that we do in fact possess this talent, albeit subtle. Researchers were quick to identify the gene responsible—CD36. It is believed that variants of this gene make us more or less sensitive to the taste of fat. You might think that being less sensitive to fat would be a good thing, but you would be wrong; it seems people who are less sensitive to the taste of fat tend to desire more of it. Scientists at Penn State University demonstrated that people with the particular low-sensitivity CD36 variant tended to favor foods with higher fat content and, as such, be at a greater risk of obesity.

How twisted a Universe to have programmed us to feel excessive comfort and satisfaction in such gluttonous behavior, leading to the deaths of so many. How cruel a Universe to have given us the intelligence to grow, understand, and improve food through science and technology, only to then punish humanity for following an instinct ingrained in them over millennia.

Technological advancements have enabled our society access to an unprecedented supply of nutritional food (for those who can afford it) as well as effective medical treatments (for those who are lucky enough to have insurance). Yet, poor dietary choices are responsible for about one in five deaths globally. Then again, premature death might be preferable to kale and wheatgrass.

Your own personal army is planning a mutiny

Inside your body right now is an army. But the army is more like something out of a fantasy novel than a real army. Think *Lord of the Rings* with monsters and giant Oliphaunts fighting alongside orcs and wizards, a crazy collection of different entities whose core job is to quickly and mercilessly destroy any invader that makes it past your outer defenses. It is not restricted by the Geneva Convention—after all, we couldn't expect such civility from the Universe. This army uses chemical and biological weapons, unconcerned with the collateral damage caused by its actions. Yes, your own personal protection system is nothing more than an assortment of cellular psychopaths, war criminals, and assassins. Just thank Gandalf it's on your side, right? If this isn't the first chapter you've read, you'll know that's not how this book works.

Every cell in your body carries a type of marker that is unique to you. It's as if each cell had one of those *Hello my name is* _____ name tags with the same name and handwriting on it. Even the cells that weren't made by your body, such as the bacteria in your gut, get a name tag, but maybe the name is different and the handwriting is the same, or the name is the same and... Forget it, we're getting bogged down in this metaphor; just know that cells welcome in your body have labels so they don't get attacked by the lunatics that make up your immune system. Usually.

Your immune system cells do go through training of sorts, but it's not at all like when you had to spend fifteen minutes watching workplace health and safety videos. In reality it's more like a boot camp. But even then, instead of getting yelled at and spit-sprayed in the face every morning, the failing recruits are just disposed of. It's intense. While part of their training includes learning how to read your cells' labels, there is so much more to it as well. Most of the cells who start this training do not finish and are destroyed at the first sign of any shortcoming. If real education bodies employed this tactic, we suspect the fail rates would be far lower, but then again so would enrollment. Is it any wonder that the graduates from the immune system training program are sadists when they watched so many of their classmates get literally pulled apart for misreading some notes?

The most common graduate is known as a neutrophil, accounting for 70 percent of your white blood cells. Neutrophils are kind of like an all-purpose soldier in your army, but rather than being tactical about how they deal with invaders, they are more like a twelve-year-old playing *Call of Duty* for the first time. They go nuts; they release a bunch of chemicals into the bloodstream, relentlessly chase down the baddies, and fire cellular bullets in every direction like it's going out of fashion. This causes your blood vessels to leak fluid, drives up your temperature, and increases your blood flow. Though it may sound bad, it actually helps fight infection, as clotting agents as well as other immune cells flood the area to join the fight against potential disease-causing factors. This effect is known as inflammation, and while helpful in the short term, prolonged inflammation can cause irreparable damage to your own body cells and really fuck you up, as you have probably seen in numerous infomercials about products with side effects even the Universe didn't think of adding to its merchandise.

Another of the graduates is called a macrophage, and they are brutal. They are larger than neutrophils, and their job is quite literally to eat the invaders. These bruisers will chase down their prey like an elephant who identifies as a wolf, until they get close enough to reach out with their trunks, grab on, and engulf the intruder, which is then hastily ripped apart by enzymes inside the cell. This whole affair is not dissimilar to the 1958 science fiction horror film *The Blob*. Disturbingly, after the macrophage disintegrates its enemy, it will then wear bits of its victim like tiny trophies, showing off its kills to its buddies in the immune system special forces.

The special forces buddies are known as lymphocytes, and perhaps you've heard of them as the B-cells and T-cells. Well, rather than being unselective maniacal murderers like neutrophils and macrophages, lymphocytes are more like serial killers or assassins with a specific target in mind. This target is determined by whatever trophy the macrophage is wearing, and this simple flex by the macrophage brings about a killing spree so violent it would make Quentin Tarantino blush. Like a shark sensing blood in the water, lymphocytes seek out anything that even slightly feels like the macrophage's trophy. If a virus is hiding out in a nerve cell, the Killer T-cells (yes, that's their real name) will just tear it to pieces. Small piece of bacterial protein floating around in the bloodstream? B-cells are going to completely coat that shit in antibodies. No evidence of the genocide orchestrated by lymphocytes is left behind except for a few B- and T-cells that stay around to sound the alarm should their target return— and if it does, then an even larger and more aggressive response is mounted. Ruthless.

But it sounds as if these cells are on *your* side. They are only out to get the villains—bacteria and viruses who we know can do so much damage. Well, that's all true, in theory. But let's not give our little defenders too much credit just yet. Rarely (but also not that rarely) some cells slip past the training academy's strict selection process and forget who the real enemy is. The rogue cells will start attacking your own body. It's almost like the Universe has watched too many movies and yearns for conflict in places where there should be none, so it forces our own army upon us.

We can't lose sight of why the Universe created these cells. It was for one and only one purpose—to kill. And millions of years of evolution, which itself is a barbarous process, has made these cells very, very good at it. Like an army, these cells like to kill together, making this business ferociously efficient. So when the name tags or labels of your body either fall off or maybe aren't read correctly, it makes for a horrific experience with no cure. It is like your own body is going through a coup d'état and the leader of the coup just wants to conduct more and more rebellions. Temporary relief from *some* symptoms is possible using drugs and is often the only option for sufferers of these so-called *autoimmune* diseases.

Scientists, with very trustworthy-looking lab coats, have suggested there are around eighty-one distinct autoimmune disorders. Most are extremely rare, so don't get on that Google medical self-diagnosis binge just yet. Although, looking at it collectively, roughly 5 percent of the population suffers from at least one of them. Most of these diseases start to show symptoms after the age of forty, clearly showing the Universe is ageist. They also disproportionately affect females, showing the Universe is also sexist. And certain autoimmune disorders present in people of color more frequently than people of European descent, showing the Universe is racist as well. Thankfully it doesn't have its own podcast.

Autoimmune disorders are also becoming more common, and at the time of writing, we don't really have a definitive answer for this increase. There are a few competing hypotheses, or it could be a combination of them, but this is cold comfort for the millions of people with type 1 diabetes, or multiple sclerosis, or arthritis, or lupus, or Crohn's disease, or vasculitis, or celiac disease, or psoriasis, or Addison's disease, or Graves' disease, or inflammatory bowel disease, or scleroderma, or aplastic anemia, or...I think you get the point. There are so many ways our immune system can turn against us and attack one part or many parts of our body. Crohn's disease, for example, can manifest anywhere in your digestive system; that means anywhere from your mouth all the way through to your anus, which if you could decide where you wanted it to show up, would be like some sort of gastrointestinal *Sophie's Choice*.

Why is such an efficient and complex structure like our immune system susceptible to such an obvious and potentially fatal flaw? Clearly there is a problem in the program, but what type of software engineer writes a line in a program that does the literal opposite of what it's meant to? Is this your first chapter? You know exactly who, or more accurately what, does this sort of thing on the regular. This fucking Universe. It has turned us into its own personal thriller/horror/sci-fi flick so it can watch internal conflict unravel the psyche of millions, enjoying a bowl of popcorn, and laughing at that scene where you ate a piece of cheese and uncontrollably shit your pants.

You are an aging mutant

Our genetics determine so much about our lives, from our obvious physical characteristics such as our hair, skin, and eye color, to the things we cannot see—like the way our internal organs function and how our body goes about breaking down certain chemicals. Genetics even has a role in our personalities. Who knows, maybe one day, saying, "Sorry, it's just who I am," will be a valid excuse for telling your boss where to stick it. And maybe, rather than being sent home with no pay, you'll be sent to a geneticist, who will send you back to the office with a doctor's certificate to show your employer, which reads: *Please excuse Alison for telling you to "shove your report up your ass," as she has the genes 5-Htr2C and 5-Htr3B. Her parents would like to apologize on her behalf.*

Our genes don't tell us only about our lives—they can also tell us about our deaths! In short, our genes can tell us what will most likely kill us. It could be cancer, addiction, or some genetic disease just waiting in the wings. All of this information is kept in the center of every cell in a biological code you likely learned about from *Jurassic Park*, DNA (or deoxyribonucleic acid if you wanted to sound smart). But it was only as recent as the early 1900s that scientists came to the conclusion that DNA was the reason you got your dad's nose. From there, many scientists began asking what we could do with this information, but according to Dr. Ian Malcolm from *Jurassic Park*, not enough were asking what we should do with it. Like, make a 2-meter-tall velociraptor.

James Watson and Francis Crick won a Nobel Prize for "acquiring" a photo from Rosalind Franklin that unlocked the secret to the structure of DNA. But don't worry, it was in the '50s, when stealing work from broads was totally acceptable. Shaped like a twisted ladder, DNA consists of just four different types of *nucleobases*, also known as base pairs, which make the rungs of the ladder. They can only pair up in a certain way—*guanine* with *cytosine*, and *adenine* with *thymine*. The side rails of the ladder are made from repeating sugar-phosphate units. Are you paying attention? There is a test at the end.

Okay, good. So, when all these things are stuck together, they make a shape that scientists refer to as a double helix. It is sobering to think that just four nucleobases have the instructions to make you and every other living thing on Earth. There is actually a fifth nucleobase known as *uracil*. It is kind of like the third member of the Apollo 11 mission—you know, not Buzz, not Neil—but the other guy. Uracil is like that. It's extremely important in making your body what it is, but not important enough that you need to mention it every time you mention DNA. By the way, his name was Michael Collins.

Even small changes in the sequence of bases can cause dramatic differences in how your body is built and how it works. For example, a change in a single nucleotide (the basic building block of DNA) can result in sickle-cell anemia. Sickle-cell anemia is a genetic disease where the red blood cells are misshapen and form a shape similar to a sickle—hence the name. This happens because the molecule that helps carry the blood, *hemoglobin*, sticks together, forming rigid rods instead of the usual globular shape. People with this abnormality cannot carry oxygen around the body, and their blood vessels have a tendency to get blocked more often. These blockages can then affect the blood supply to the brain, the heart... Look, essentially what we're saying is a tiny change in your DNA can mess with your heart and your brain. And if there is one thing Dorothy's friends taught us, it's that you can get away with missing one, but definitely not both.

There are over thirty billion base pairings in your DNA. They form long noodle-like strands called chromatin, which wind up tight into forty-six chromosomes—twenty-three from each of your biological parents. The good news is, most of the mutations that occur have no noticeable effect on your body. How could they? We don't actually know what the point of most of those thirty billion base pairs are. What we do know is that the number of chromosomes has no bearing on the complexity of the life-form carrying them. A fruit fly has eight chromosomes, a pistachio has thirty chromosomes, and there is a type of fern with fourteen hundred. Even the humble potato has two more chromosomes than you.

Chimpanzees, gorillas, orangutans, and all great apes also

happen to have forty-eight chromosomes, two more than humans. However, scientists discovered something interesting when looking at the structure of our DNA in comparison to our closest living cousins. A few million years ago, before humans came to be in their current forms, something happened to a common ancestor of ours. This event was so significant that it would change the future of the planet. It could even be argued this event was the catalyst for an entirely new geological period, comparable to the time an asteroid slammed into the Earth and wiped out most of the dinosaurs. Two of the chromosomes belonging to our ancestors became fused together, and a mutant ape was born. Usually, when something like this happens, the mutant fails to grow and develop, and will die while still in the womb, but not this one. It survived and led to us. This improbable survival story is the reason why our chimp cousins have an extra pair of chromosomes more than us and also why heroic survival movies against all the odds are so universally loved... we assume.

This fused chromosome is still in all of us today. How do we know this? Science, of course. Science has allowed us to map out in exquisite detail the structure of our chromosomes. Unfortunately, when we looked, our chromosomes didn't have labels on them. So we came up with a very clever naming system: chromosome 1, Steve, chromosome 3, chromosome 4... Ah! Just checking if you were still paying attention. There is no Steve chromosome. But our second chromosome is a little different from the rest. While most chromosomes have repeating DNA sequences at their ends called *telomeres*, Steve has telomeres in the *middle* of it.

Telomeres are like protective caps on the ends of our chromosomes. They are kind of like that little plastic sheath on the tips of your shoelaces. The tip of your shoelace is called an aglet, by the way—the more you know! Anyway, the telomeres act as a buffer in protecting the sequence of DNA and (usually) prevent the chromosomes from fusing together. Non-telomeric DNA is subject to degradation, but telomeric DNA is specially structured so it is not as vulnerable to wear and tear. So, we should be thankful to these telomeres located on the ends of our chromosomes.

Every time a cell divides, a little bit of that telomeric DNA is lost. So, telomeres are somewhat of a biological clock, in that they can determine the life span of an organism. Worse still, all the fun stuff like eating junk food, drinking alcohol, and using drugs can accelerate the shortening of these telomeres. This means as you get older, the protective ends of your chromosomes are a little less effective and your DNA is a little more likely to degrade. Scientists are still unclear if shorter telomeres are a result of the aging process or are the cause of it.

When the telomere length gets too small, the cell it's contained in is removed by your immune system. But don't worry. It goes to a nicer place, a happy farm with all the other cells. Just kidding; it's actually euthanized and dropped from your body with the rest of the trash. This programmed death allows for your body to remove damaged, cancerous, or virus-affected cells. In cancer cells, an enzyme called telomerase assists in extending the life of the telomeres. An oversupply of the enzyme telomerase could be a good indication of cancer. In fact, researchers have actually been able to keep cells alive for far longer in the lab by using telomerase. Interestingly, lobsters have large amounts of telomerase in their cells and they don't display the effects of aging like the rest of us. Scientists studying aging have found that, on average, people with longer telomeres will live five years longer than those with shorter telomeres. It won't be long before you start seeing commercials for telomere lengthening serums on questionable sites. *Want to live like a lobster? Grow your telomeres by 5 inches with these pills!*

But no need to stress if you have shorter than average telomeres—it's not the size of the boat; it's the motion in the ocean... Oh wait, it is totally about length in this case. Anyway, this is just one of many factors that will kill you. It's really a group-effort type of situation. With things such as gender, lifestyle, income, age, and now genetics all coming together, you can certainly find the right mix to put yourself six feet under.

When looking at the function of telomeres and the other factors influencing our life expectancy, one can't help but think of the 2011 sci-fi movie *In Time*. The film is set in the year 2169 and people stop aging when they are twenty-five. Once a person reaches this age, a timer appears on their arm. This results in time being a universal currency—meaning it can be traded, both fairly or otherwise. From this information alone, you don't need a spoiler alert to guess what happens next. The more powerful a person is, the more time they have. Like all good dystopian fiction, there is some moral truth to this. The lifestyle choices you make can often determine the length of your time on this Earth. It's no surprise that, statistically speaking, the wealthier you are, the longer you will live. In fact, a paper published in 2016 found that the difference in life expectancy between the richest 1 percent and the poorest 1 percent in the United States was 14.6 years. Not quite enough to make for a good plotline, but still, more evidence that time and money are transferable. So if you have some free time, hit us up. All interesting trades considered.

So, here you are—a mutant ape that is aware of the fragility of life; an ape that has the potential to understand the fundamental mechanisms of vitality, the Universe, and how to extend your time on this Earth; an ape that is intelligent enough to know how to minimize the destruction of your own genome; an ape that can easily avoid *diseases of affluence* such as heart disease and type 2 diabetes. Here you are—a mutant ape, scratching your ass, ignoring everything you just read.

There are tiny assassins everywhere

Our species has walked the Earth for around 300,000 years, and during this time, an estimated 117 billion of us have lived and/or died. That number is unfathomably large. In fact, if we went back in time 117 billion seconds, we might find ourselves enslaved by Egyptians building pyramids to honor dead pharaohs. But pick up just a handful of the particularly rich Nile River soil, and you would have about this number of bacteria in your palm. Now imagine how many bacteria are strewn across all the land, rivers, and oceans on this planet. We hope you have your hand sanitizer ready!

Maybe you're thinking this isn't a fair comparison. After all, a bacterium is an entire organism made up of just one cell, while humans are made of somewhere between thirty and forty trillion cells. All right, fine, Rain Man, let's do some more cellular accounting. For every human cell that makes up your body, there are about 1.3 bacterial cells. That means even on your own self, there are around forty-eight *trillion* bacteria that call *you* home. Makes you feel a little less lonely, doesn't it? Or maybe not. Shall we keep rolling with these numbers?

At the time of writing this sentence, the global human population is just over 8 billion people. That means there are around 384 sextillion or 384,000,000,000,000,000,000,000 bacteria collectively living in or on human bodies right now. Talk about overpopulation! Let us remind you that this number is *only* for humans. This doesn't include the bacteria that live on or in other animals and plants, nor those that are swimming in oceans and rivers, live in soil, populate every surface in your house, or are just simply floating in the air around you. We are yet to find an environment on Earth that isn't inhabited by some form of bacteria. And by the way, scientists have managed to catalog only an estimated 1 percent of the bacteria species that exist. In a nutshell, there's a lot of fucking bacteria out there.

So why doesn't the whole planet look like a giant petri dish or some sandwich found in the bottom of a bag from your last vacation? Why don't you look like some sort of swamp thing all the time, instead of just at the end of a particularly loose camping trip? The simple answer is that animal and plant cells are on average about twenty times larger than bacteria cells. The detailed answer is a little more complicated than that, but you have time, right? *Pfft*, of course you do... So bacteria in our bodies move around a bit and generally don't congregate in large numbers, unlike the cells that make up our tissues, such as skin and muscles. In short, they are tiny, even by cell standards, and they don't stick around for a very long period of time. In fact, if you were to remove all of the bacteria in your body, leaving only human cells and their various secretions, even the highest estimates say you would only lose about a kilogram of mass. For our American readers, that's about the mass of 150 bullets or a little more than two pounds. In any case, while bacteria are high in number, there just isn't very much to them.

You may be wondering though how we can go through life swarming with bacteria and yet manage to avoid being permanently bedridden with infection. Well, most of the bacteria are actually neutral to us, meaning they are like Luxembourg: small, forgettable, and anything they do does very little to affect us in any way. Some bacteria are actually beneficial; they help us digest food, maintain the health of our skin, and even support our immune system, though you are unlikely to find them advertised as ingredients in health and beauty products. *Maybe she's born with it. Maybe it's a kilo of bacteria.* With the magic of modern medicine, you can get a bunch of them shoved up your ass. And it's not just for the enjoyment of it; poor gut health is continually being linked to mental illness and a range of other ailments.

But problems can arise once a critical number of bacteria is passed. Even these otherwise harmless bacteria can go on to cause havoc, especially when released *en masse* in the wrong area. You can think of them like college students—often benign, occasionally helpful, but when they congregate in high numbers in a Mexican resort town and start reproducing, disease is assured to follow.

Nevertheless, we have a long-running war with bacteria that predates recorded history. It predates the existence of man and probably even the evolution of fish. What is shown in our recorded history is a largely one-sided scoreboard with the microbes resembling the Harlem Globetrotters and humans the Washington Generals, the bacteria firmly on top with a score in the billions, repeatedly ravaging human populations with diseases such as cholera, tetanus, and tuberculosis. And much like the real Globetrotters versus Generals, the game is heavily rigged against us. The famed Black Death that wiped out about one-third of the population of Europe was caused by the bacteria *Yersinia pestis*. Then there's syphilis and gonorrhea, two more bacterial diseases that may have a relatively low death count but certainly know how to hit a human where it hurts. Jarringly, we didn't even know these microscopic war machines existed until about 350 years ago, and even then we didn't recognize them as the cause of so much suffering for another 150 years. Essentially the Universe had dumped invisible assassins in every conceivable location and then given them a few billion years to prepare for the arrival of their targets.

There have been multiple attempts to fight back throughout human history, using naturally antibacterial chemicals like silver and copper, or processes like inoculation. But usually, these efforts fall embarrassingly short of the death quotas filled by periodic epidemics and outbreaks. For a brief window, it looks like we might be finally getting one up on these malicious microbes. The invention of antibiotics, the adoption of widespread vaccination, improvements in hygiene, and environmental management practices have substantially reduced the death rates of infectious diseases over the last hundred years or so... Well, this is at least true in developed countries; tuberculosis and cholera still feature high up in the causes-of-death list in developing countries. But don't worry; that overpriced shirt you bought at Live Aid made a real difference.

This apparent turning of the tide may be short-lived with antibiotic-resistant bacteria becoming increasingly common and many citizens of developed countries deciding they know better than doctors and epidemiologists and refusing vaccines. This is problematic because of a couple of quirks given to bacteria by the Universe: *fast reproduction rates* and *horizontal gene transfer*. You know all about the former—winky face, eggplant—but the latter is way more interesting.

Horizontal gene transfer is a fancy term scientists use to describe how bacteria can exchange bits of their DNA with other bacteria, even between completely different species. Imagine being able to take a copy of Angelina Jolie's eye genes, using it to reshape your own eyes, and then finally nailing that smoky eye look. Or maybe incorporate some horse genes into your own DNA before kickstarting your porn career. Bacteria do this kind of thing regularly, and this process helps quickly spread antibiotic resistance through a population.

Given enough resources, some species of bacteria can double their population every twenty minutes. This means a single bacterium could become many billions of bacteria in the space of twenty-four hours, rendering you incapacitated and covered in your own feces before you have time to get your lazy ass to the doctor and beg for the vaccine your upper-middle-class monkey brain thought was unnecessary. This scenario rarely happens because you have an immune system, but even that can go wrong (see Reason 13).

The ongoing arms race against bacteria is not likely to end

soon, and this war on disease is being fought on a number of fronts. If you're worried about bacteria after reading everything up to this point, don't be. Let us tell you about viruses instead.

Viruses are even smaller than bacteria, only about one-twentieth their size. They are also far more numerous. Remember when we discussed how many bacteria call your body home? Well, there are about tenfold that number of viruses on and in you. Viruses can only reproduce by hijacking a host cell, but unlike bacteria, they cannot do it on their own. This means when bacteria invade your body, they survive and reproduce outside of your cells. But when a virus invades, it gets inside your cells and uses them like a sex doll, making thousands of copies of itself before your cells burst open, resembling over-filled balloons and releasing the newly formed viruses, which go on infecting other cells. Feeling violated? You should.

Viruses are about as close as it comes to a pure biological machine. They are the microbe equivalent of the Terminator—extremely robust, quick to evolve, and often elusive to conventional medicine, especially since they hide out in your body's cells. However, most viruses are only able to invade one type of cell in a particular type of organism. In other words, like the Terminator, they are extremely specific, focused solely on finding their *Sarah Connor*.

Occasionally a virus will make the jump between species, and all hell breaks loose for a time (COVID-19, anyone?). But regardless of who or what species we blame for an epidemic or pandemic, one thing remains true: viruses are the most likely pathogen to bring about the apocalypse.

It seems likely that viruses have existed for at least as long as life itself, and they have caused great suffering for humans as long as we have walked the Earth, from smallpox and polio to influenza and genital warts. Even recent history is dotted with examples of viruses that emerged causing economic and social chaos, only to fade away, leaving small pockets of the disease and lots of angst; think bird flu, swine flu, Zika, Ebola, AIDS, COVID-19, monkeypox, and, probably by the time you are reading this, something like Alphaomegacron Elephantpox.

The Universe has been able to produce not one but two groups of microbes able to inhabit literally every nook and cranny on the planet. At the same time, it gave them a reproductive potential that makes even the horniest rabbits look about as fertile as Lance Armstrong's excised testicle. But surely we can't end this chapter on a cancer joke.

Literally everything is poisonous

In January 2000, the British physician Harold (Fred) Shipman was found guilty of murdering fifteen people over three decades. It was believed that this number could have been as high as 250 people—with his youngest victim possibly being just four years of age. His preferred poison? Diamorphine—better known by its street name, heroin. His efforts would award him the name of Dr. Death. Another Englishman, Graham Young, while other fourteen-year-old boys were spending an inordinate amount of time cleaning their rooms with the doors locked, spent his time poisoning friends and family by lacing their drinks with antimony. Standard serial killer naming conventions granted him the name the Teacup Poisoner.

Then there is Jim Jones, whose legacy lives on in the saying, "Don't drink the Kool-Aid." He is infamous for the deaths of more than nine hundred people who were members of his cult. He spent all his money on expensive cyanide, which meant he actually had to mix it with the cheap knockoff Flavor Aid instead of Kool-Aid to kill them. These are horrible examples of human beings using a terrifying modus operandi to execute people. Odorless and tasteless, these poisons readily slip past the natural poison detection systems our bodies have developed. But perhaps the most harrowing aspect of the weapon of choice for these people is the minuscule amount required to wreak serious havoc. But we urge you, don't follow these examples—just go with the Kool-Aid itself, which is supercheap on Amazon Prime, or go to Costco should you need it in bulk.

To determine how dangerous a chemical is, scientists use a measure called the LD50. LD stands for *lethal dose* and the 50 refers to 50 percent of a population. Basically, to determine this number, scientists administer a known amount of a suspected poison to a bunch of lab animals and see what amount of the chemical is required for 50 percent of the sample population to die. Yeah, we know, this all sounds pretty fucked up, but science isn't free. Once they have established the LD50 for mice, rats, rabbits, or whatever poor creatures find themselves in these types of studies, they can extrapolate the data for humans. It would be far more accurate and efficient to conduct these studies using human specimens, but volunteers for these kinds of studies are extremely difficult to come by, except maybe on Craigslist.

While this all sounds horrific, because it kind of is, it is also important to realize that these data are used for all kinds of medical treatments, environmental health studies, and industrial chemical hazard reduction. Now is the time we should all take a moment to remember and honor all those lab animals that unwillingly sacrificed their lives to make ours safer... Okay, that's long enough.

Those chemicals mentioned in the first paragraph have quite low LD50s, which means the murderers who used them didn't need much to carry out their dastardly deeds. Heroin has an LD50 of about 20 mg/kg, which means an amount about the size of a sugar cube could kill an 80 kg person, unless that person is a regular user and has developed a tolerance for it; then it'll take something more like four or five sugar cubes to do the job. Antimony is less deadly, requiring a couple of egg-sized portions to be eaten before an 80 kg person kicks it. But, it'd be a horrible way to die—burning stomach pains, severe vomiting, and watery diarrhea—which goes some way to explaining why Graham Young's death count was low but the degree of suffering he caused was so high. Cyanide, on the other hand, is incredibly toxic: for an 80 kg person to kick the bucket, only a few grains of that sugar cube need to be stirred into their Kool-Aid. However, these aren't even close to the most toxic poisons floating around.

Ricin can be made from certain beans and is of similar tox-icity to cyanide if eaten, but if inhaled, requires only a thou-sandth of this dose. Batrachotoxin is the stuff secreted by the poison arrow frogs from the Amazon, and a dose the size of only a couple grains of sugar is plenty enough to kill most people. It is also a good argument for not licking toads. Maitotoxin can be found in poorly prepared shellfish, especially if it was har-vested following algal blooms. The dose required to kill an 80 kg person is too small to see with the naked eye—only about 10 micrograms. Take that sugar cube analogy we've been using and divide it into one hundred thousand equally sized grains, and you would only need one of them. However, toxicologists agree that the most deadly known substance on Earth is botulinum. It is so toxic that to kill an 80 kg person you'd only require about 80 *nano*grams. That's the size of our sugar cube divided into 100 million pieces, and even then you wouldn't need a whole piece to do the job. Despite the toxicity, many people readily inject it into their face, with botulinum being the active ingredient in the wrinkle treatment we call Botox. Yes, countless Californians are shooting the deadliest toxin we know directly into their face, but hey, you look five years younger.

To add insult to injury, most of these deadly chemicals are produced by other life-forms as a literal example of overkill. These highly dangerous chemicals are not uncommon. Cyanide can be found in lima beans, almonds, apples, and many other common foods, although the amount is far below the lethal dose. Nevertheless, when you think about deadly venoms produced by life-forms, you probably think about the snakes of Australia, or the spiders of Australia, or the jellyfish of Australia. But even the most toxic of these have an LD50 dose similar to cyanide—hardly the worst thing out there. Although, to be fair, some of these animals deliver the equivalent of about one hundred doses in a single bite.

We know what you're thinking: *I've avoided these poisons all my life, and I have no intention of traveling to Australia.* Well first, deaths from animal bites are extremely rare down under, although drop bear fatalities are at an all-time high according to the locals. Second, most of these poisons were only discovered after they had already entered some poor soul's body, causing severe reactions or death that could not be explained by infection or other known causes. This means there may be even worse poisons out there that have yet to present themselves in some back-alley soup kitchen or maybe have not yet evolved in some over-the-top evolutionary arms race.

Now you're probably thinking, *It's still fine, I'll just be sure to cook all my food properly and stay away from jungles, the Russian secret service, and dodgy-looking food trucks.* But we have only discussed the worst poisons out there. The implication from using the LD50 measurement is that it means *everything* is a poison. The real difference between these chemicals and some others you are more comfortable with, like whatever is in the dust from the Doritos you just licked off your fingers, is the *dosage.* Think about the chemicals that are essential to your life. We're not talking about the caffeine in your coffee, although you should know that a AAA-size battery amount of pure caffeine would probably kill you. We're talking about the oxygen you breathe, the water you drink, the sugar that fuels your energy. These chemicals are also deadly at the wrong dosage.

Oxygen concentrations above about 50 percent, apart from making spontaneous combustion a real possibility, will lead to irreversible lung damage after about twenty-four hours of exposure. Indeed, in another experiment that could be considered an ethical minefield, baby mice that were exposed to pure oxygen suffered brain damage and other complications you've probably read about in the fine print of daytime television ads. By the way, if you were thinking oxygen was purely a good thing, you may want to flick back to Reason 2. The next time an oxygen mask drops down in front of you, maybe just test it out on someone else first—the armrest hog in the reclined seat is a good option.

Water, which seems like something you'd desperately turn to as a cleanser of all things dirty and poisonous, is really no better. For example, if you drank more than six milk cartons of water in a single sitting—as some moronic health guru has surely suggested—it would have disastrous effects on the body. The result is a condition called hyponatremia, wherein the sodium levels in the body become so dilute that the brain swells, causing seizures and coma and eventually leading to death. This actually happened to a twenty-seven-year old California woman in 2007 who was taking part in a radio contest. The contest involved participants drinking lots of water and then holding in their "wee" to win a Nintendo Wii. Get it? Which marketing genius came up with that? Anyway, the woman left the competition empty-handed and full-bladdered before complaining of a headache before heading home, where she was later found dead from hyponatremia. A jury awarded her family $16.5 million in damages, which is the equivalent of about sixty-six thousand Nintendo Wiis.

And what about the sugar we've been talking about throughout this chapter? How much sugar do you need to eat before it kills you? Well, it's about 2 kg worth, which is the size of the standard sugar bag you get from the grocery store and probably only slightly less than your average Cinnabon. Although, as long as you don't plan to eat the whole thing, you should be okay.

The fact that the Universe has made literally everything a poison really speaks to its character, but it is more nuanced than it seems. Numerical values siphoned off the back of the mass murder of lab animals do not really give you an accurate picture of how deadly these chemicals are. How the chemical enters your body, either inhaled, ingested, or injected, will affect its toxicity. Also, your previous small and sublethal exposures will build a tolerance to a particular poison. The would-be assassins of Russian monk Rasputin found this out after multiple attempts to poison him failed. They ultimately went for the less cunning—but more efficient—bullet to the head, which we don't recommend trying to build up a tolerance to.

REASON

Sex sucks

Undeniably sex, or the potential for it, is a great motivator for much of the behavior we and our fellow earthlings exhibit. Many of the smells and sounds around us are all an attempt by living things to copulate. From a bird's song at dusk, to the aroma of fall roses, to the grinding of two intoxicated strangers on a dance floor, these are just various organisms trying to mingle their DNA one way or another. But have you ever wondered where sex originated from, or better yet, the absurd destinations to which evolution has taken it? Unsurprisingly, the answers seem to match up with what we'd expect from a pervasively deviant universe.

Our best evidence suggests that life, albeit in very primitive form, first appeared around 3.5 billion years ago. The first eukaryotic cells—the more complicated cells that gave rise to the cellular structures we see in basically everything that isn't bacteria—appeared around 2.7 billion years ago. It wasn't until 700 million years later that we find the first evidence of sexual reproduction in some complicated and horny cell blobs called *protists*. At present, 99.9 percent of all eukaryotes, at least on occasion, reproduce sexually. But the idea of pleasurable sex is only a relatively new thing—and, for many people, remains a nonexistent phenomenon. In fact, it has only been in the last 600 million years that the first animals developed neurons *capable* of feeling any kind of pleasure. But sex isn't just about pleasure; it has been shown to produce something known to scientists as "offspring." We understand this may come as a surprise to many readers in their early twenties. But this is important for the survival of the species, as sexual reproduction can allow for greater variation of the genetics within a population. Unlike *asexual* reproduction—where an organism makes a genetic copy of itself—sex has evolved to become a billion-dollar online industry. Be that as it may, some would argue that humans at least attempting asexual reproduction is not without its profitability.

The evolution of sex has followed a similar path to that of the pornography industry. At first, it was very straightforward—tab A into slot B kind of stuff. But at some point, the Universe's desires could no longer be satisfied by a simple exchange of genetic material between two individuals. It has evolved penises that resemble a medieval mace, orgies that Larry Flynt would find uncomfortable, or mating that makes BDSM look like foreplay with a feather. With all the damage that these intromittent organs can cause, it's no wonder that sex is such a great vector for disease.

Let us introduce you to the *antechinus*, a small marsupial found in Australia and the Universe's answer to a teenage boy who just got a computer put in his bedroom. These nocturnal rodent-like creatures live out the first ten months of their lives feasting on spiders, centipedes, and frogs, much like other marsupials of a similar size. But while most marsupials eat to sustain themselves through winter, the antechinus eats to sustain itself for the first weeks of spring. What happens in the first weeks of spring you ask? A. Gigantic. Marathon. Orgy.

When the moment arrives, a male can copulate with a single female for an amazing fourteen hours—which is about thirteen hours and fifty-eight minutes longer than most human men. And if that doesn't make our male readers feel somewhat inadequate, when the antechinus wraps things up with that female, he moves straight on to the next one, then the next one, then the next one, then the next one... You probably get the point. But like a male antechinus, we aren't done yet. This sex fest can last for days to weeks. The little love machine continues to go for broke, even though his supply of baby gravy has dried up. He exhausts himself and eventually goes blind, just like your grandmother warned you could happen. His blood is coursing with so much testosterone he could maybe even place third in the Tour de France. Eventually, he starts to fall apart...literally. His fur falls out, he has internal bleeding, and a failing immune system leaves him gangrenous and septic. But he still doesn't stop! He continues searching for the next lover, but largely to no avail. At this stage, most of the females—themselves tired, weary, and exhausted—have gone into hiding. But being blind means the males often can't tell anyway, and there are reported cases of antechinuses fornicating with broom heads, believing the bristles to be a female. Eventually, all the males end up looking less like cute marsupials and more like a hairless cat that's barely survived being hit by a Toyota Prius. Diseased, exhausted, and blind, the curtain closes on the antechinus's life. His legacy will live on with his offspring...and in someone's broom head that they accidentally left outside overnight.

If the tormented tale of the antechinus isn't enough to convince you that the Universe has deluded us into thinking sex is pleasurable, let's talk about cat penises. If you thought your cute little kitty cat had it easy, how wrong you were. A cat's penis is best described as what a dildo would be like if it were constructed on the set of a Mad Max film: spiked, unpleasant, rough, and, in short, painful—much like how we would expect a conversation with Mel Gibson to go.

The spikes on the cat's penis are known as *penile spines*, and they are as distressing as they sound. The spines serve to scrape the walls of the female's vagina and, as such, stimulate ovulation. Unlike a ribbed condom, these penile spines are not designed for pleasure. In fact, the act of sexual intercourse can cause serious damage to the female cat's reproductive organs, which probably explains why little Mittens was making the neighbor's cat produce those horrible sounds. And if you think that's the worst of it, you're in for a treat. One of the worst sexual reproductive acts happens in your very own bed—when you are in it!

Imagine lying in bed fast asleep, but next to you something is about to go down—something so foul, so unimaginably dark and sadistic, it makes Jack the Ripper look like Santa Claus. As you lay there asleep, an unwelcome pair of intruders have made their way into your bed. First, the male climbs on the female. Seems fine so far. But then, using a sharp object, he stabs the female in her abdomen. The female, left with a hole in her side, just lies there. With the same instrument used to stab the female, he ejaculates in the wound. No, your neighbors aren't from some teen horror flick—you have bedbugs.

Unlike the female, the male bedbug is lucky, as he will live on. In creatures such as spiders and praying mantises, it is the female who gets the last laugh. In the case of the praying mantis, the female steals more than just the male's heart; she also steals his head. The male, decapitated, continues to persevere. Headless or not, he is determined to get the job done. So much so, that a male mantis can continue to mate for hours after his head has been bitten off. At least with spiders, the females usually let the male finish before eating them (most of the time).

Resembling some talent show, the male peacock spider, which gets its name from his impressive abdomen that resembles the tail plumage of a peacock, dances for his survival. Rather than a red buzzer being pressed, if the male peacock spider does not dance vigorously enough or pay adequate attention to the female, she will, well, kill him. And if the male peacock spider thinks he can just rehash his old dance moves, he is sorely mistaken. Recent research has indicated that once having already mated, the female peacock spiders become harder to impress. Let's just be thankful the rest of the animal kingdom doesn't have their own version of the Internet.

These are just some of the many examples of how the Universe has set up a system whereby if a species is to survive, they must risk either being eaten, stabbed, scarred, or diseased, or continue to update their repertoire of dance moves. Remember this the next time you tell someone that having sex is *totally natural,*" because so is having your head ripped off after it.

REASON

We're probably already halfway to our extinction

We live in a random Universe. Some things are less likely, like a large asteroid hitting the Earth next Wednesday. Other things are far more likely, like a large asteroid hitting the Earth within the next billion years. What makes one of these things more likely than the other? Well, the timescale.

Many people will live out their lives having fire insurance, even though it is doubtful their house will be burned to the ground—unless you live in Australia… Actually, let's just assume that when we talk about freakish weather events, poisonous animals, or anything involving a patch of land that humans shouldn't inhabit, Australia is the exception. Alas, probability is a reliable way to assess and mitigate risk, and with this in mind, enter the *Doomsday Argument*.

The Doomsday Argument uses probability to predict the number of humans that will exist in the future and where our species will "max out," if you will. Oh, and if you think that this is some abstract idea, this reasoning was instrumental in predicting the value for events such as the expansion of the Universe and even the fall of the Berlin Wall. The Doomsday Argument just takes things further. The Doomsday Argument suggests that there exists two scenarios awaiting humanity: Quick Doom, whereby the entire scope of humanity is capped at 200 billion humans (with about 100 billion already living or having lived), or Later Doom, whereby humanity will successfully advance our civilization to have upward of 200 trillion humans. Why these numbers specifically? Let's start with the idea of *Quick Doom*.

Trying to figure out how many humans have ever lived is a rather difficult question to answer. There are a number of things that one needs to take into consideration. For example, we don't *really* know how long humans have been walking the Earth. No, we're not validating creationism—we'd like to take this moment to thank our creationist readers for getting this far in the book, though—we are highlighting a key feature of evolution, its graduality.

Evolution is for the most part a slow, imperceptible process. That is to say, thinking that some sort of ancestral primate gave birth to what we now consider a "modern" human, is like looking for the exact location that yellow become green on a rainbow; in reality it's a gradual transition that is very hard to decipher where one thing ends and another begins. For reasons such as this, we will never know the *exact* date at which the first *humans* walked the Earth because evolution just doesn't happen that way.

Most people without fanatical religious views about the history of man would acknowledge that real evidence suggests humans, or *Homo sapiens*, have been around for about 300,000 years. If you want to be a real stickler and are looking for the anatomically indistinguishable hamburger-eating, gas-guzzling, beer-drinking *Homo sapiens* that stares back at you in the mirror, you only need to go back 200,000, or you could look at the dude sitting left of you on the subway.

Now that we have an estimate of the amount of time that modern humans have existed, we need to understand the population trends of past civilizations. This is a bit more difficult to determine because most of these civilizations didn't keep records on account of not having paper or a desire for administrative work. For the others, many of the records that were kept have been wiped out for various reasons, like natural disasters or covering up genocides. The very few records we do have capture only a small part of our 200,000-year history, and most are about as trustworthy as a gas station hot dog. Even so, we can use other techniques, such as fossil examination or carbon dating, to make up for all these shortcomings. All things considered, we estimate that about 100 billion people have existed on Earth so far.

The Quick Doom argument says that statistically speaking, the 100 billion people who have ever lived represent around half of all the people who will ever live. This idea might seem odd at first, but imagine having the test scores from every single high school student who ever sat for a nationwide test. If you randomly picked a single student's score, it is most likely to be somewhere near the average for no other reason than most students' scores are near the average. If graphs are your thing, let us just say the single life suits you, don't listen to the haters. But, for your chart-oriented mind, we are saying that the test score is likely near the center of the *bell curve*. Sure, the score you picked could be an extremely low score or an extremely high score, but the chances of picking one of those are immensely unlikely.

Now replace these test scores with the entire past and future population of the human race. And the randomly selected score from the list is...you. Feeling special yet? Well, you're at least as special as the other 8 billion or so other test scores walking around the planet right now. As one of the chosen, you are likely close to the average, which means there are just as many humans ahead of you as there are behind you. That is, there are another 100 billion or so humans left to go before the last one.

If the statistics have convinced you so far, you are probably wondering when humanity will hit the 200 billion mark. Well, there are a few more things that we need to consider, such as changing birth rates as developing nations transition into more advanced economies, war, famine, disease, and so on. However, if our current efforts are anything to go by, it could approach sooner than most of us would like.

Humans bred their way to a population of one billion people in 1800. Not bad, considering large-scale soap production wouldn't occur for another 50 years. We hit a population of two billion in 1927, despite hiccups from World War I and the Spanish Influenza pandemic. We had doubled the population again to four billion by 1975 following the baby boom, and rather counterintuitively, the invention of the contraceptive pill. In 2022 we surpassed eight billion people, and if trends continue, we should reach the grand finale of our species in about 760 years. It's a pretty poor effort considering most comparable species last on Earth for around a million years or so.

That's depressing, but what about the better Doom, a.k.a. *Later Doom*? Ah, yes...the better of the two Dooms. Later Doom is the scenario where humanity lives far longer, thriving in seemingly endless prosperity. In this outcome, humans move beyond Earth, colonizing the solar system and beyond, exploiting resources on countless other planets and moons. Our population explodes, matched only by our limitless desire for expansion, and we become masters of the galaxy. Sounds great, right? Well, get ready for some more bad news... We have another analogy. And if having to read another analogy hasn't disappointed you enough, it describes why the *Later Doom* scenario is far less likely to occur.

Imagine you are a contestant on a game show. There are two identical boxes: one contains ten balls (numbered 1–10) and one contains 500 balls (numbered 1–500). The rules are simple—choose the box with fewer balls and win $1,000,000. You cannot look into the boxes, but you are allowed to reach into one of the boxes and pull out a single ball and read the number.

Unless you're an idiot—then again, you did agree to go on this stupid game show—you should realize that if you get a number between 11 and 500, then you should select the other box. As luck would have it though, you reach into one of the boxes and pull out a ball with the number 7 on it. Shit. This means it could still be either box. We told you this game show was stupid. But wait, if you really think about it, it is far more likely that it came from the box with fewer balls. Why? There is a 10 percent chance of pulling a 7 from one box, while only a 0.02 percent chance from the other. You like those odds.

Unfortunately, this analogy suggests that the chances of us

being at the beginning of humanity's existence are very low. In other words, we are the balls and the fact that we got picked probably means there aren't many balls to pick from. So although you would be likely to take home the prize in our hypothetical game show, the real prize you get from this analogy is existential dread! Congratulations! It's likely to last longer than $1 million in this economy anyway.

The Universe will have the last say as to when, how, and, even if humanity perishes. We are just mere passengers with an illusion that we are somewhat special. Yes, it does *appear* that from our current observations we are unique, but everything from our star, solar system, and galaxy screams mediocrity. Quick Doom or Later Doom, doom is inevitable. Math doesn't lie, and the Universe loves math.

PART IV

THIS PLANET ISN'T THAT GREAT

We should probably find somewhere else to ruin.

REASON

Our best friends are also our enemies

Statistically speaking, you are likely a pet owner. Rates of pet ownership are as high as 70 percent in the United States. Pet ownership averaged across all of Asia and the Oceanic region is also quite high at around 60 percent. Some nations in Central and South America reach household pet ownership rates of 80 percent. More than half of the world's population has some sort of pet, with dogs being the most popular. And boy do we love our canine companions, with some reports saying 98 percent of dog owners consider their furry friend a fully-fledged member of the family.

However, data from the United States estimates that one in seventy-three people will be bitten by a dog, with higher chances for the very young. And what about the chances of being sexually abused by an overly impassioned dog? Well, we don't have data on it, but it is probably way higher. You probably didn't imagine all the things that designer dog was meant for, did you? Statistics show that about 25 percent of dog attacks requiring hospitalization occur in the home. This means one-quarter of serious dog attacks were from a dog that was at home and most likely involved a person known to the dog or its owner. Man's best friend, huh?

Interestingly, Rover researchers have examined the brains of dogs while they are shown pictures of their owners, and as best as we can tell, our dogs really do adore us. Their brain activity resembles many of the patterns and produces many of the chemicals that we would associate with love. When this experiment is repeated with our second most popular pet, cats, we get zilch. Does this mean that your favorite feline doesn't love you? That you are just somewhere to grab a quick snack on their way to killing some local birdlife before having loud, distressing sex with the tabby down the street at 3 a.m.? Not necessarily. But it does suggest that your pussycat sees you more like that dorky kid from elementary school who always had the best snacks.

Don't be mistaken, there are a raft of well-recognized benefits to pet ownership, from improvements in mental health to increased life expectancy. But there is a Noah's ark of well-recognized germs and diseases that dogs and cats quite happily carry on and in themselves that frequently infect humans, including the almost indestructible superbug antibiotic-resistant *Staphylococcus aureus* (see Reason 15). Maybe it's time poochy had a bath? The chances of experiencing significant harm from these guys may not be zero, but it is low. For almost twenty thousand years, dogs and cats have been a regular part of human society. We have selectively bred them to our desires, and sharing our homes for so long has allowed us to coevolve with them. But what about some of the other animals with whom we share our world?

Something like 60 percent of emerging infectious diseases originated from an animal. Diseases that evolve to jump the species barrier from animal to human are called *zoonotic diseases* and boast an impressive lineup of well-recognized players, including HIV/AIDS, Ebola, Zika, and of course, COVID-19. In fact, if we look at just the last thirty years of data, around thirty brand-new microbes were discovered to cause disease in humans, and 75 percent of those came from an animal. Still consider yourself an animal lover? Make sure you wear protection.

Before we get lambasted by PETA, or accused of starting some sort of crusade against all creatures great and small, let's make one thing clear about this chapter: we don't hate animals. Except wasps—fuck those things. The Universe has done horrible things to the animal family tree. It has turned many of them into drunks, racists, and unemployed conspiracy theorists. Sure, there are some that we don't mind hanging around, but they are firmly in the minority when it comes to our closest living relatives.

Often when people think about animals to avoid in nature, they instinctively snap to the Australian environment, and it's fair enough in one sense. Australia has twenty of the top twenty-five most venomous snakes in the world; it is also home to several of the most venomous spiders, most venomous jellyfish, and most venomous octopus. Even male platypuses have venomous barbs on their hind legs. It is indeed a place you don't want to be bitten by something, but only about two people a year die in Australia from snake bites and fewer than twenty people die from any kind of envenomation annually. Would you rather be bitten by a Taipan and be in a race against the clock to get some anti-venom, or be mauled by a lion and left bleeding out in the African savanna, picked apart by vultures, hyenas, and jackals? It's like deciding whether you want to compete in the Olympics (without steroids) or the Hunger Games.

Let's linger in Africa for a moment longer. There are several species of big cats there, and even the herbivores are terrifying. Hippos, elephants, and buffalos are responsible for around twelve hundred deaths a year. Ever heard of someone getting trampled by a kangaroo? Didn't think so—they'll fuck you up if you try to box them, though. In Africa, there are also gorillas and chimpanzees, hyenas, wild dogs, and countless more animals that could turn a casual hike into an extreme sport. Africa also has a completely different assortment of venomous creatures, including its own versions of terrifying snakes. But we're talking about Australia and Africa, the two most objectively terrifying continents on the planet, right? Well maybe, let's think about this a little more.

North America has bears, really big ones that get angry when they can't find a *pic a nic* basket after a long nap; they kill about two people per year. There are also cougars, wolves, buffalo, and moose. Moose directly injure more people than grizzly and black bears combined and cause multiple deaths indirectly due to car accidents associated with them crossing highways, although only the Canadian moose feel *soorrey aboot* it. Moose are big, like *really* big. Oh, and North America also has venomous snakes.

South America has jaguars, caiman, anacondas, poisonous frogs, the goliath bird-eating spider, a venomous centipede that grows as long as a thirty-centimeter ruler, and also some venomous snakes. Asia has the Komodo dragon, leopards, tigers, Asian elephants, a lethal snail, and also some venomous snakes. In Europe, you can find wolves, bears, wild boars, and also some venomous snakes. Even Antarctica has leopard seals. No venomous snakes, though...

In all of these places, we have not even come close to naming all the animals that could kill you, or at the very least, ruin your pants. And all these continental land masses make up only a fraction of the Earth's animal habitats. Venture into the oceans and waterways and you can also add on sharks, electric eels, piranhas, crocodiles, orca, giant squid, stonefish, and a bunch of venomous sea snakes. In other words, a very sizable portion of the animals that we share the planet with are able to kill us without even trying. If an animal decides that they simply don't want us around (and who could blame them?), then our fleshy exterior and lack of horns, talons, sharp teeth, scales, or pretty much any defense is easily penetrated. We know that our American readers are probably thinking guns are the answer right about now. But even guns won't stop many animals, as the Australian Army corps found out after surrendering to a large herd of emu in the Great Emu War of 1932.

In the right situation, most of the animals mentioned above would instill fear in even the most hardened criminal. Perhaps even Batman should reconsider his choice of animal mantle. But all of the animals named above in this chapter don't even come close to the death count of the most deadly animal on the planet. Humans... Just kidding, we wouldn't do you like that—it's the mosquito. Still, we doubt Mosquitoman is likely to become anything more than a straight-to-DVD, B-grade Horror/Sci-Fi movie. But what this 2005 movie, directed by Tibor Takács (seriously, look it up), lacks in popularity is made up for by the death toll of the real miniature menaces. Maybe Tibor Takács was onto something and there is still time for Marvel or DC to snap up the rights to this character.

We did say earlier in the chapter that we don't hate animals but, seriously, fuck these guys. Mosquitoes are responsible for over 750,000 deaths per year. Largely due to the diseases they carry, in particular, malaria. But they also carry Zika, elephantiasis, yellow fever, dengue fever, Ross River virus, West Nile virus, Japanese encephalitis, chikungunya, myxomatosis, and many, many more, specific to the particular region a mosquito species is found. Wherever you go, mosquitoes are the same—jerk-offs with hypodermic needles strapped to their face, prepping to inject you with pestilence while you sleep.

The point is that no matter where you go, there are animals around so scary looking that they make Willem Dafoe look like he belongs on the side of a bus in a family dentist advertisement. There is always something to bite, poison, trample, maul, scratch, infect, or hump a human. It's not their fault—it's a consequence of their conditioning, and to be fair probably human ignorance. Be as it may, what sort of universe creates a planet swarming with life-forms and then decides to put venomous snakes everywhere?

REASON

The ground could give way at any moment

Earth is really big. We know if you've read some of the space chapters, it may not seem like it, but for this chapter, it is massive. And when we say Earth, we don't just mean the surface. Most of Earth is as unpredictable as Brad Pitt at a Tanzanian orphanage. But the crust, a paltry 0.45 percent of the Earth by mass, is livable, solid, and unmoving, more like Doug Pitt, Brad's younger brother, businessman and philanthropist who is also the United Nations Tanzanian Goodwill Ambassador... Wait a minute. Surely not? Anyway, it's clear for anyone to see which is more stable. Or is it?

An extremely scant amount of the Earth is actually survivable, which is a depressing enough fact that it ought to be a bona fide reason to hate the Universe. Luckily, with foresight, we've covered it in Reason 22. But let's talk about the more livable places on Earth, which are made from rock that slides around so slowly even David Attenborough's film crew couldn't capture it with a time-lapse camera. Occasionally, these sliding rocks can be surprisingly nimble, leaping and bounding like a ninety-six-year-old Sir David ascending in a cherry picker to really nail that aerial shot. Yes, even the usually reliable and stable area on the surface of our ball in space can crack and unleash on an unsuspecting group of people. You don't want to see old Dave when he finally snaps.

The Earth's surface is basically a jigsaw puzzle with a bunch of different-sized pieces. But instead of being pieced together on a sturdy flat table, it floats on top of a globe of liquid that flows like uncooked brownie batter. And instead of the pieces fitting nicely together, they bash up against each other; sometimes one sinks below another only to become part of the brownie batter. Is this making you hungry? Hangry? Horngry? Be right back.

Thanks for waiting. See how for you, that felt like no time at all, while for us, it felt like it took forever to make an organic, vegan toasted sandwich just to feel better about eating a bag of nacho cheese corn chips and ten store bought chocolate brownies in front of a few episodes of *Blue Planet*. Well, that's how the tectonic plates feel about the entire history of our Genus. The continents have remained relatively unchanged since the first *Homo erectus* started to migrate north from modern-day Ethiopia around two million years ago. So yeah, it's slow going with this continental drift stuff.

In fact, these jigsaw pieces move so slowly that even the fastest of them drags along at only about two-thirds the length of your cell phone each year. (It's the Pacific plate, by the way. But that's an oceanic plate that no one lives on or cares about.) The much bigger, slower Eurasian plate, which includes countries like Kosovo and Tibet, moves at a blistering speed of about two finger widths per year. That means over the course of your lifetime, Europe and Asia would have moved southwest by about as far as you can step. But, given enough time, this shift will change the climate patterns, create land bridges between continents, and alter the ocean currents. Since plate tectonics are so slow, human civilization will probably not live to see these effects. Alas, a generation of us thought it would be cool to see them anyway and decided to burn a bunch of old carbon to change the climate. In any case, don't let this slow movement of totally independent countries deceive you. While the *average* movement of the planet may seem minuscule, every now and then, the Universe fucks us.

Think back to when you were a kid and were given your first gobstopper. You were an observant kid, so you could tell this was going to be a hard candy, but you trusted your newly grown molars as they had been cracking rock candy like quail eggs for months. You put the gobstopper into your mouth and bit down. It really was hard, even harder than you expected, but your jaw muscles had only just begun. You squeezed down on the candy, and the resistance in your jaw muscles increased, but nothing happened. You kept squeezing, and before long, your jaw felt a burning that can only precede fatigue. You gave it one last hard squeeze, and finally...*crack!* You chipped your tooth. Now that would have cost your parents a pretty penny unless, of course, you are British. In any case, this is similar to what happens at the boundary between Earth's tectonic plates. Pressure builds and builds until it's released, resulting in lots of broken shit, large repair bills, and frantic reading of the fine print on your insurance contract.

You may have heard of the Richter scale, which measures the size of earthquakes. It is calculated using a few of those jiggly needles drawing lines on paper called *seismographs*. By measuring the size of the wiggle and the time it was detected in a few different places, scientists can determine where, when, and how big an earthquake was. Imagine standing in a completely dark room with your closest friends and hearing someone fart. If everyone points in the direction they heard the fart, you can determine the origin. But if you and your friends were flatulence seismographs, you'd also be able to tell us the precise decibels of the fart and if it was protein- or fiber-based. But, unlike a good bottom burp, there is no way for us to control when an earthquake will occur.

We are remarkably bad at earthquake Whac-A-Mole. We can't even predict whether earthquakes will occur within the span of decades, let alone how big they will be. The *best* we can do is provide a few seconds—that's right, *a few seconds*—of warning to areas miles away from the epicenter. And, of course, that is *after* the earthquake has occurred. Imagine answering the door and Lindsay Lohan is standing there saying, "I'm going to punch you in the face in three, two, one..." Except after an earthquake, instead of getting punched in the face, you have a pile of rubble where your house once stood and no celebrity to hang a lawsuit on.

We don't want to alarm anyone, but you could be in an earthquake right now! Don't worry though, as it's probably one of the minor earthquakes that can't be felt. Thankfully, these are the most common ones, and they end up on the Richter scale with magnitudes less than 1. Earth experiences an earthquake with a magnitude of 6 on the Richter scale about one hundred times a year. Now, going from 1 to 6 doesn't seem that bad—except that the Richter scale is logarithmic, which means a one-point increase on the scale equates to a tenfold increase in power. If you did the calculation in your head, you'd know this means a magnitude 6 earthquake is *one hundred thousand* times stronger than a magnitude 1 earthquake. The 2011 earthquake that all but destroyed Christchurch in New Zealand was around a magnitude 6. However, when it comes to earthquakes, much like the excuse of every geologist you've ever slept with, size isn't everything.

The proximity of the earthquake to the surface of the Earth and its duration also play a role in the amount of damage caused. The Christchurch earthquake was so damaging because it occurred close to the surface, directly below the city. The 2004 Indian Ocean earthquake that triggered a tsunami on Boxing Day measured 9 on the Richter scale and lasted for ten minutes. This resulted in the sea rising to a height of almost thirty-eight stacked meerkats and inundated surrounding islands, killing about a quarter of a million people and at least thirty-seven meerkats.

Most earthquakes last a pitiful twenty seconds, but like every geologist you've ever slept with has no doubt said, "Twenty seconds is plenty of time to make the bed rock." Following an earthquake, land can be raised (or dropped, depending on your perspective) by several meters. The earthquake can also move land sideways, snapping roads, railways, and other long stuff that is supposed to stay long. Earthquakes move land to a more comfortable place, from the Earth's perspective anyway, and there the land will remain until the next quake, which could be days, months, years, or decades away. As best we can tell, the

Universe just wants us to know it's coming for us, but it doesn't want us to know when.

Whether by karma or just the Universe's queer notion of bad luck, let's say you are unfortunate enough to be caught in an earthquake. What do you have to look forward to? First, the buildings, rocks, and basically any toppling objects that are taller than a few meerkats standing on the head of David Attenborough are a real threat. So, you decide the best course of action is to run out into the open paddock behind your house because you're a rancher for some reason. Bad decision. Sure, you have the land, but the income is not very stable. Besides, what good is all that land when earthquakes cause *soil liquefaction*? As the ground starts shaking, dirt turns into viscous mud, and sand becomes quicksand. Anything standing on these loose surfaces will start to sink. Entire villages have been lost to soil liquefaction as buildings topple, not directly due to the shaking, but because the foundations are swallowed into the Earth like a scene reminiscent of *Star Wars, Dune,* or *Tremors*—minus the giant man-eating worms and a screaming Kevin Bacon, of course.

So what should you do if you live in an earthquake-prone area and the shaking begins? The best you could do is get under something like a table and, just like with every geologist you've ever slept with, hope it'll all be over soon. Earthquakes are just another example of how the gradual, almost peaceful movements of the Universe have often been shoved aside in favor of violence with no warning, no consideration of guilt, and little chance of escape for anyone but the top meerkat.

REASON

The Earth is covered in explosive pimples

Imagine if the only nutritious part of an apple was the skin, while the flesh, seeds, and stem weren't just benign, they were toxic (which is still sort of true; see Reason 16). That's almost the equivalent thickness of the survivable area on Earth, a thin crust floating atop thick, viscous instant death. But at least with our hypothetical apple, the noxious insides don't bubble to the surface as they do on Earth, or explode violently, destroying a huge chunk of the apple and changing the air surrounding it for centuries. Of course, we are talking about volcanoes!

Volcanoes have long been idolized by ancient cultures as gods and by science teachers as a craft activity. They command respect and we have given it. We have romanticized them in our art, and they are ubiquitous across every civilization from the ancient Minoans to modern parents doing their child's science fair project the night before it's due. And what have we received in return for all this respect? Death and destruction—also some pretty amazing science, but mostly death and destruction.

Some volcanoes are very inconvenient. They erupt, pumping dust, ash, and other pollutants into the atmosphere. Take, for example, the Icelandic volcano Eyjafjallajökull (we'd love to know what word you said in your head just now). It erupted on and off for a few months in 2010, causing the cancellation of about one hundred thousand flights. But this isn't a reason to hate volcanoes. If Eyjafjallajökull was a person, you might not like them, since they made you cancel your trip to Rome, but you surely wouldn't harbor a deep, bitter hatred about them. If you disagree, we know of some people from nearby Pompeii who might tell you cancelling a trip to Rome is the least of your worries.

You see, Eyjafjallajökull is a real baby volcano that we already had an eye on and kind of knew how it was likely to behave. There are lots that remain beyond our vision, like the ones on the bottom of the ocean or beneath the Antarctic ice. Who knows when or how much they'll erupt, but it'll probably be just another inconvenience. Even more concerning are some of the other ones that we are watching, and they are much, much bigger. Should they erupt, forget about cancelling your plans: cancel the whole planet.

Volcanoes have played a role in three of the past five mass extinctions, yet asteroids seem to take all of the glory for global exterminations. Volcanoes were probably the major contributor to the end of the Triassic extinction, which wiped out 50 percent of species on Earth. Now that's a decent reason to hate them. Whole civilizations have been exterminated by volcanoes at speeds of which sixteenth-century European colonizers would be jealous. There were large populations from Indonesia, South America, and the Mediterranean lost to us because of volcanic eruptions—and that is just a few within the last twelve thousand years of recorded history. Who knows how many tribes of geniuses or superhuman genes were wiped from the face of the Earth because an angry mountain spewed as if it went too hard at a Golden Corral buffet.

You might be surprised to find out that these mostly inconvenient but sometimes deadly Earth pimples form because the Earth's skin is constantly in motion. The thickness of the Earth's crust is less like one of those perfect apples on the weekly produce flyer and varies more like one of those deformed apples no one wants, except apparently when they order their groceries online. At its thinnest, the floor beneath our feet is only as thick as seven thousand cocks stacked atop one another. We're not sure what image you have in your mind right now, but we hope it is illustrative of just how thin the flooring on our home planet is. We're glad it's there though, since beneath the floor, rocks melt and behave like cold honey. The thin part of the crust is where the tectonic plates are sliding past each other. When a continental plate slides over an oceanic plate, it generates a lot of heat, and some of the less dense molten rock bubbles to the surface. If the bubbling goes on for long enough, an Earth pimple forms, which is awkward because the Earth is certainly middle-aged by now and should really be done with acne.

Calling volcanoes *Earth pimples* is a wildly inaccurate description of how a volcano works, but it does allow us to draw some interesting analogies. Not all pimples, whether on the Earth or your face, are the same. Some are big and so sore you can feel the pressure build up in them until one day...*pop!* All over the mirror. Some are small and clustered and are more likely to weep slowly over a week. And some just seem to keep poking up in the same spot only to burst, subside, and then return months or years later. We've likely ruined the romance for volcanoes at this point, unless of course, you're one of those people who loves watching pimples pop and are just now realizing you also like geology.

Volcanoes appear more frequently in certain places, like pimples that show up on your face and ass. However, hot spots beneath the surface of the Earth also mean they can rise up from pretty much anywhere, like a pimple in your armpit. Hawaii is known for its beaches, waves, and weather. But active volcanoes are also part of the appeal for millions of tourists, generating well over $150 million per year for Hawaii. Here's the shocker for geology buffs (and those way too invested in this chapter): these idyllic islands are nowhere near a plate boundary! In fact, the Hawaiian volcanoes are more like the Icelandic one we met earlier... What was its name again? EI-re-plostaffja... Forget it, let's just call it Greg. Greg and his buddies are majestic, peaceful, and predictably only throw a few tantrums every now and then. Be like Greg.

A famous volcano you probably shouldn't be like is the one in Yellowstone National Park. To be fair, it is indeed pretty, peaceful, and has a geyser on it named Old Faithful for being so predictable that scientists can guess when it will erupt more accurately than United Airlines can schedule its departures. It would be reasonable to assume that you don't need to worry about this volcano, since Yellowstone attracts over four million tourists who collectively spend about $500 million every year. But Yellowstone is no ordinary volcano; it is a *supervolcano*. Supervolcanoes are exactly what they sound like—volcanoes whose eruptions are as violent as it gets, wiping entire ecosystems from the area and changing global weather patterns for centuries. These aren't Earth pimples; these are more like full-blown ass boils, guaranteed to go viral on the Dr. Pimple Popper YouTube channel, and there ain't no cream for them.

The last time Yellowstone erupted was about 640,000 years ago, and scientists don't think it will happen again for at least another million, if ever. But the chances are not zero, and Yellowstone is just one of about a dozen supervolcanoes identified on Earth. For a volcano to spew out hot lava, it has to be full of stuff beforehand. Yes, it's as obvious as it sounds. We don't have an eye on what's inside Yellowstone, but if it were full and the pressure built up enough, say goodbye to Wyoming, Montana, and Idaho, as all of those states would be buried in ash as deep as fifteen reams of paper or about fifty copies of this book, assuming you are reading the paperback version. The loss of that many books would be a tragedy. Whether anyone would notice these states were missing is debatable.

After the initial explosion, which would be bad enough, it would begin raining down glass and rock across North America on what would become the worst possible day to try and catch snowflakes on your tongue. The global climate would cool over several years, but don't think for a minute an exploding volcano is a reliable fix for global warming. The last time any supervolcano erupted was about seventy-five thousand years ago, and it triggered a winter that even George R. R. Martin would describe as *unreasonably long*. The Earth remained in the cold season for eight years following the *Toba catastrophe,* which is believed to have almost wiped out humanity. We mean, it wouldn't be anywhere near the first time a volcano took out a species, or several, but when it's our species, it makes it a little more personal. Though, can we really blame the Universe for trying?

So please, don't romanticize volcanoes. Just casually check them out as a tourist or watch a PBS documentary about them. You could even pretend to be a geologist for a day. They are the sexiest scientists, after all, rocking their khaki shorts, socks-with-sandals combos, and greasy hair leading into a knotted ponytail. The confidence to rock this look alone is so arousing... mmm... Anyway, volcanoes are just a manifestation of this asshole Universe, who not only decided we can't live *in* the planet but likes to send up periodic reminders of why we can barely live *on* it. Volcanoes are not to be worshiped but rather acknowledged for what they really are: violent manifestations of the hostility this Universe has for life. We are totally powerless to do anything about them except to run away.

When Mount St. Helens erupted in 1980, that's precisely what locals did—well, most of them. The worst thing about this particular eruption is that instead of erupting up, Mount St. Helens erupted sideways, blowing hot ash, mud, and rock across the land like a passed-out frat boy vomiting on his friend's pillow. Local badass Harry R. Truman and his sixteen cats were vaporized after telling evacuation crews and journalists he wasn't leaving because he was "part of the fucking mountain." Well, he's not wrong now. Another stayer, volcanologist David

A. Johnston, was stationed about ten kilometers away and was tasked with monitoring the eruption for the USGS, an undertaking he unquestionably succeeded in, sending a final radio transmission: "Vancouver, Vancouver, this is it!" Unfortunately, his remains were never found. Love the Universe, and what do you get back? A coffin-less funeral and a brief mention in a book about how shit the Universe is.

Even the livable places are shit

Think back to that time you took a wrong turn, ended up in New Jersey, and thought to yourself *This is a shithole; I couldn't possibly live here.* Well, you weren't wrong about that, but almost everywhere is New Jersey in a way. In other parts of this book, we talked about how inhospitable the Universe can be, but have you ever stopped to think about just how inhospitable the Earth itself is?

Living on Earth is kind of like installing an Olympic-sized swimming pool in your backyard but never leaving the top step at the shallow end. The Earth's surface area is around 510 million square kilometers or 71.5 *billion* football pitches, or about the size of fifty-two United States of Americas, or sixty-four Australias, or thirty Russias, or 3.2 million Liechtensteins. If you wanted to cover the surface of the Earth in thick protective skin, it would require twenty-nine *trillion* Indian elephants...and they're endangered. Why would you want to do that?

But 71 percent of Earth's surface is covered by the oceans, which is great for *some* fish and aquatic life, but even then, about 90 percent of marine creatures live only in the top 300 meters. Considering the ocean can reach depths of about eleven kilometers, or 4,500 Indian elephants (with skin) stacked on top of each other, this is a paltry amount of habitable area compared to the total area in our oceans. The *average* depth of all the world's oceans is about 3.5 kilometers. You could walk that distance in less than an hour, so why then are most living things only hanging around in the shallowest 8 percent? Once again, it comes back to physics and the Universe's utter contempt for making life easy.

Light is really only able to penetrate to a *maximum* depth of about 330 stacked Indian elephants, but everything past about sixty-seven stacked Indian elephants' worth of depth is essentially in perpetual night with only a small amount of blue light reaching down there. This means no aquatic plants or algae grow here, and as a result of this, there are no herbivores and very little oxygen. The few organisms that do dwell in this zone permanently depend on a constant rain of *detritus*, which is essentially dead animals and shit, to provide nutrients that allow them to survive. Imagine if the only food you could eat was just broken up bits of dead things... Wait a minute. Well, at least we're not eating shit. This decay rain continues to sink to the bottom, so the animals here have to be quick too, or else they might miss out.

As you go deeper and deeper, the pressure around you gets higher and higher. At a depth of five hundred stacked Indian elephants, the pressure is over 2,000 PSI. PSI is the standard measurement of pressure and stands for *pounds per square inch*. As a point of reference, a basketball is 8 PSI while your blood pressure is about 2 PSI—higher if you've invited indignant relatives over for dinner. This means at 1.5 kilometers down, the pressure is about 140 times greater than at the surface. This pressure would certainly collapse your lungs and potentially break some ribs were you to somehow swim that deep down without the use of a submarine. At the bottom of the Mariana Trench, the deepest point in the ocean, the pressure is a whopping 17,000 PSI. Many people think that a human would fold up or explode from a squeeze this size. But instead, water would likely force its way into all your orifices, and you would end up swelling like you were a ShamWow that Vince Offer wiped over a wine-soaked countertop. But wait! There's more. We should clarify that we are only assuming this *would* happen; no one has actually done the experiment.

The opposite happens whenever we try to bring up creatures from the deep, and this time, we *have* done the experiment. By bringing them to the surface, the fluid that once filled their cavities oozes out of their orifices, and they expand like balloons. Take, for example, the blobfish. In its natural environment, it resembles a spiny Vin Diesel (on set). But when drawn to the surface, it looks like a sad and sloppy Danny DeVito (on set).

But let's climb out of the ocean to where the remaining 29 percent of Earth's surface is land. It would seem there's plenty of room for us to live on, but most of it is actually uninhabitable. If it is not a frozen wasteland, it is a hot sandy desert devoid of any water or nutrients that could sustain a diversity of life. Don't even get us started on the mountainous regions, which are steep, hard to walk up, and at altitudes where the atmosphere stays so cold that water freezes and most plants won't grow. Though adventurous humans have risked life and limbs to climb the tallest peaks in order to dump their trash and take selfies, unfortunately, they didn't stay.

Even when we look at the locations that would be theoretically habitable for humans, most are inaccessible; think the deepest jungles in Borneo or the thousands of small islands dotted throughout the oceans. Unless you like the idea of retreating to a commune to live in isolation, they are also not very practical. Many of these places are already inhabited by plants and animals that would not appreciate us encroaching on their homes (surprise, nature hates us—see Reason 19). Besides, living in isolated communities has its own drawbacks, such as starvation, limited access to health care, lack of Wi-Fi, and an unhealthy attraction to relatives not so far removed.

So here we are, stuck to the surface of a spinning rock. Most of it, we can't even live on without either killing ourselves or fucking up something else. Oh, and don't bother trying to leave, because the Universe is even more forbidding off the rock than on it. But stop... You're thinking in only two dimensions. What if, instead of going across the surface, we decide to go down into the soil and rock, and build a new society there, well below the moles and earthworms and even the deepest plant roots? We won't have to worry about disturbing nature, there's plenty of groundwater to be found, the temperatures are far more stable, and we can funnel light using mirrors! Problem solved! Well, not quite.

Much like the oceans, there is a sweet spot for underground living, and as you go deeper and deeper, the temperature gets hotter and hotter. At about five hundred stacked Indian elephants down, less than half the depth of the ocean, the temperature underground can sit at a sultry 45°C all year round. The deepest mine in the world, about twelve hundred stacked Indian elephants deep, reaches temperatures of 60°C. There are a number of towns that do reside just under the surface of the Earth. Coober Pedy in Australia or Berber caves in Tunisia are prime examples. But pretty much every place that utilizes underground housing exists in some sort of extreme environment. Again, take Coober Pedy, where surface temperatures range from 6°C to 50°C, and the local real estate agents are forced to describe it as something for everyone.

The deepest humans have ever dug into the Earth's crust is about twelve kilometers. At this depth, the temperature is over 200°C. Now you should feel more comfortable telling Ken to stop bitching about the air-conditioning at the office—it could be much worse. But even the impressive distance of twelve kilometers is only about 0.2 percent of the thickness of the Earth, and the entirety of Earth's crust is about twice this thick. We live on a crust that has a thickness equivalent to a layer of tin foil wrapped around a tennis ball, and a gentle whack with a racket (or perhaps an asteroid—see Reason 32) could easily break it open.

Beneath the crust is the mantle. It is 3,000 kilometers, or about one million stacked Indian elephants (last one, we swear), thick of melted rock, swirling around at temperatures in excess of 1,500°C. We think it's a pretty safe bet that there is no chance of living there either.

Unfortunately, it just gets worse as you get deeper. In the core, it's hotter, denser, and probably has a bunch of radiation and electrical currents you wouldn't want to contend with. Basically, the inside of the Earth makes the bottom of the ocean, even with all its violating water, sound like paradise. Don't believe what some movies tell you—no one is ever going to the center of the Earth.

But why bother going down when we could go up? Maybe we could make better use of the atmosphere. Floating cities far above the surface of the Earth, anyone? Well, that would be fine if the oxygen content wasn't so low, the radiation exposure so high, and the fact that the wind blows at a relatively constant 90 km/h. Sounds great...if you want to be cancer riddled and gasping for air while living in a never-ending hurricane. So, basically, we're stuck where we are. Bunker down and learn to love your home, as it could be a lot worse. At least we're not being forced to live off a constant drizzle of poop particles.

Yet again, whenever it seems like the Universe has thrown us a bone, providing a nice planet to live with all our needs provided, by digging just a little deeper, we realize just how stingy the Universe really is. Even the only known harborer of life in the Universe is 99.9 percent inhospitable to it. Carl Sagan asked us to look out into the cosmos and marvel at the enormity of it. But what ought to be a humble exercise in humility is really just a stark reminder that the Universe has imprisoned us in a tiny, isolated corner of a hostile hellscape.

By the way, an Indian elephant is about 3.1 meters tall.

REASON

Most water will kill you

Representatives in the New Zealand federal parliament have twice tried to introduce bills that ban the use of the chemical *dihydrogen monoxide*—once in 2001 and another time in 2007. In 2004, city officials of Aliso Viejo in Orange County discussed banning foam cups after learning that *dihydrogen monoxide* was used heavily in the production process. This seemingly troublesome and toxic chemical is better known by its common name, *water*.

But even having that knowledge doesn't seem to stop people from trying to exploit our inherent fear of chemicals. A waterfront park in Kentucky tried to deter people from bathing in a public fountain by erecting a sign reading *DANGER! WATER CONTAINS HIGH LEVELS OF HYDROGEN—KEEP OUT!* Now, most people—except non-swimmers—would probably not see *water* as something to be worried about, let alone something that can kill you. But water's crystal-clear public persona only came about fairly recently and is a consequence of good marketing and good science. Even now, in some places, a small sip of water could mean an untimely and very messy death.

You've probably heard that to stay healthy and hydrated, each person should drink eight glasses of water per day. That's mostly bullshit. Many self-styled lifestyle coaches mindlessly vomit out facts and figures about how "water makes up 75 percent of your body mass" before trying to sell you some sort of special health-promoting water, like *alkaline water, chlorophyll water,* or *gemstone-activated water* (yes, this is a thing). These products are only good for one thing—curing your heavy-wallet syndrome. Maybe your parents banned you from having sugary drinks, instead handing you a glass of water, saying those other drinks will "rot your teeth," which is half true. Perhaps you have read somewhere, specifically in Reason 29, that one of the things astrobiologists are looking for in their search for alien life is liquid water. Whatever *you* know about water, there is one thing that everyone knows—*water is essential to life.* Well, sort of... Terms and conditions apply.

First, let's think about what water actually is. You may know it by its chemical formula, H_2O. Basically, that's telling you that water is a molecule made of two hydrogen atoms and one oxygen atom. Picture it like Mickey Mouse's head, where the ears are the hydrogen atoms and the main part of the head is the oxygen atom. But please don't hold the picture for too long, or else Disney may sue. This shape gives water the impressive ability to dissolve all kinds of stuff in it, like salts, minerals, some vitamins, and, you know, other substances... We don't judge. This means we can use water to move things around in our body, which is really important when trying to get the good stuff in and the bad stuff out. You are no doubt intimately aware of the main way water leaves your body—from your fun bits! Yep, assuming your kidneys are normal, your pee is about 95 percent water, with the other 5 percent made up of salts, bacteria, and other rubbish your body needs to get rid of posthaste.

The structure of water is super important in making sure your body is in *balance*. No, we're not talking about the same kind of balance those aforementioned health coaches are using to try and swindle you out of money. We're talking about *homeostasis*. It's not something you can buy at the drugstore, but we're certain you could find someone, somewhere trying to sell it. Homeostasis is the science jargon for how your body controls your temperature, blood pressure, salt, and oxygen concentrations. In short, homeostasis is all the processes that keep you alive—and it's all powered by water. So drink it up! But wait, isn't this chapter about how water will kill you? We hope you are sitting down for this—preferably not in a bathtub. Water is actually one of the most deadly chemicals in the Universe. How's that for a plot twist?

What we think of as "water" isn't really how water exists in the real world. Most of the water on Earth is in the oceans with concentrations of salt so high that it would kill us in a matter of days, if not sooner, were we to drink it. In order to remove the extra salt in seawater, we would need to pee out more water than we consumed. So our inability to drink from the ocean already rules out about 97 percent of the available water on Earth as a source of hydration. Of the remaining 3 percent of the water on Earth, most of it is solid ice, locked up in glaciers.

While humans have been doing their best to melt these glaciers by burning every ounce of coal they can get their hands on, removing water from giant ice slabs for us to drink presents a whole new host of problems. From moving the ice to where it is needed, to the energy required to make it into potable water, to the environmental consequences of having a giant chunk of ice sitting somewhere it's not supposed to be, it's possible, just not feasible. Another big chunk of the fresh water on the planet is found deep underground—unimaginatively named *groundwater*. While we can pump up *some* of it, as you would expect, it is dirty and requires a lot of filtration, and don't think for a second that that *As Seen On TV* plastic kitchen sink attachment you bought is going to turn mud into water. Water treatment requires several different processes using a vast array of chemicals to ensure your water is safe enough for you to drink. So again, groundwater is another largely unusable source of fresh water.

Of all the water that exists on the surface of Earth, only a minute amount is available for human use. So that's bad. When you think about the size and depth of the oceans as well as all the glaciers and groundwater that exists, 0.3 percent is actually still quite a lot of water. So that's good. But hold up—we said that water is "available" for human "use." What does that mean? Spoiler: it doesn't mean we can drink it. You may have forgotten that we aren't the only form of life on Earth, and there are many other earthlings that also depend on water. And no, we aren't talking about lovable Spot, the neighbor's dog who you let drink from the garden hose. We are talking about *microscopic* life.

There are some nasty characters in the microscopic world— some of which are certainly not your friends and can really ruin your day, your week, your month, or even your year. That is to say, some of these bastards could just straight up kill you—or at the very least make you shit your pants. We didn't even know these little combatants existed until relatively recently. You can read about it more in Reason 15 if you have the stomach for it. So even when you consider the proportionately very small amount of water we could *potentially* use, a lot of it has already been claimed by some formidable foes. Many developed nations were able to fight back once we knew what we were up against in the battle for available water. As mentioned above, we have created processes to kill off these microbes and purify the water using chemicals, heat, clever machines, plastic bottles, some glue, a label, a picture of Fiji, shrink-wrap, pallets, shipping containers, long-haul trucks, warehouses, non-unionized labor, and a self-checkout to bring you a product a thousand times more expensive than what you can get out of the tap. And that's capitalism, baby!

We can now make 100 percent pure water, unblemished by germs, toxins, or high salt concentrations. The only problem is that our bodies also don't handle pure water very well either. Given enough time and volume, *deionized* water will actually flood your cells and can cause them to burst, while at the same time stripping essential salts and minerals from your body. Luckily, pure water doesn't really exist in nature, despite what bottled water companies tell you. It's like bottling the atmosphere from a Scottish isle and labeling it as "pure air from Shetland," then selling it to working-class Beijing residents who cannot afford to travel there. Yes, this kind of stuff really happens, and this "pure air" is actually a mix of the atmosphere, salty water vapor, wildflower pollen, and the cumulative stench of thousands of tiny horses.

Now we know that drinkable water is extremely scarce and easily contaminated. Collectively, we spend hundreds of billions of dollars to ensure we have clean water safe from germs and potential diseases. So with this in mind, what do humans do to this incredibly valuable resource? They shit in it. Both literally and metaphorically. Developing nations often lack the infrastructure to separate drinking water and wastewater. Consequently, people in these countries are forced to wash, bathe, and defecate in practically the same place. In developed nations the water that runs into our kitchens for drinking comes through the same pipes and delivers the same drinkable water into toilets. Companies are collectively fined billions of dollars each year for polluting waterways, many of which serve as sources of drinking water for groups of people. It certainly seems like drinking water is undervalued wherever you go.

All of these facts prompt a very important question—why don't we make better use of the water available? Come on—we're not that lazy. We have been trying for centuries. Supplying consistently clean, potable water for the masses has been one of the greatest and longest running challenges in all of human history. Many of us in the developed world take it for granted that we can stroll into our kitchen, turn a faucet, and get some of that crystal clear, hydrating life juice—and practically for free! Meanwhile, in the developing nations of the world, the water will kill you. Also, if you live in Flint, Michigan, or Woburn, Massachusetts, or the other 610 drinking water sources across forty-three states that have elevated levels of toxic compounds. Actually, don't worry. You don't need to look for the contaminated water; it'll find you.

REASON

The entirety of nature is against you

Next time you leave the house, look around at the seemingly endless beauty of life and nature. Take in the colorful flowers with intricate details, the smell of rain clinging to lush green grass, the mesmerizing patterns of starlings as they swirl in front of the setting Sun, or even the serenity of the forest on a hike. As you do, take a deep breath (but not too much—see Reason 2). Nature is fucking lit. But be sure not to look too closely, because then you'll be forced to see the pure evil lurking beneath that alluring facade.

Those flowers? Plant genitals that take advantage of starving pollinators by using them to carry sperm to eager females. That rain smell? It's called petrichor and is a volatile mixture of bacteria and their secretions. Those starlings? They're just playing a safety-in-numbers game, trying to avoid being eaten or freezing, by huddling. And tropical rainforests? Oh boy, it's ruthless there, basically just a kill-or-be-killed situation where the serenity you feel is probably more like the tension of a Mexican standoff. So why does this remorseless, cold-blooded, and savage Universe present itself with such a pleasing pretense? It wants you to fuck, which is not as cool as it sounds (see Reason 17).

We'd like to say it didn't always used to be this way, that the brutality came from some sort of trauma or that the need to struggle for survival was just "tough love." But alas, from the very beginning, it seems if you want to get by in this world, you'd better be willing to let others die, or at the very least, take advantage of them. It's no coincidence that every major leap in evolution seems to have immediately preceded or followed huge extinctions.

Earth was once a hellscape dotted with hot, acidic, soupy cesspools covered in sheets of slimy microbial mats, strewn across a very volcanically active surface. If you are imagining Detroit, Michigan, you are getting close, but there were also no plants, no animals, not even cockroaches, mushrooms, or the remains of a once-thriving automobile industry. Just some weird bacteria and an atmosphere made up of fart gasses and the products of rot. To these first life-forms, it was paradise. One day, a special type of bacteria emerged that could use these gasses in another way; it made oxygen, which slowly filled the atmosphere. Good for us now, but oxygen was incredibly poisonous to the first life-forms, and most of them died off.

Fast-forward a billion years or so, and Earth is oxygen-rich and covered in all kinds of different bacteria. Some of them still work like the genocidal oxygen puker, some have even learned to use the oxygen and make a ton of energy from it, while some are just bigger and eat smaller ones, stealing their energy. One day, a big sucker, instead of eating the little guy, just trapped them, enslaving the bacteria inside its walls for all eternity, thereby granting itself an almost limitless supply of energy to grow and reproduce without ever having to work again. Sort of like a hedge fund manager living off the money made by actual workers while simultaneously complaining about how busy they are.

Next, the cells tried to fight back against the Universe's demand for death or subjugation and decided to come together as a community and share resources instead of fighting over them. But evolution is in it for the long haul, and it had played this game before. All that happened is more and more communities popped up, became more and more specialized, and the arms race went from unicellular to multicellular. What happened to all these characters from the first season of multicellular life? Wiped out. Eighty-six percent of them. Then fish and crabs took over, but like some thief in the night, algae stole most of the oxygen from the oceans. Seventy-five percent of life dies. But again, life bounces back, then *bam!* Ninety-five percent of it disappears in a period paleontologists refer to as the Great Dying. Cue the age of the dinosaurs. Well, we all know how that ended—in a deep fryer breaded with eleven secret herbs and spices.

The point is, the environment wants you dead—not just you personally, but life itself. Even the defining mechanism that produces new types of life, evolution by means of natural selection, is bent on destroying its own creations by forcing them to fight in a Thunderdome arena we call an *ecosystem*, which in itself is a pretty shitty battleground (see Reason 4). The harshness of life was summed up by Charles Darwin in his wildly sought after, chart-busting fifth edition of *On the Origin of Species* when he commandeered the phrase "survival of the fittest." This doesn't mean that only the CrossFit bros will make it through the next population bottleneck while the rest of us are destined for eradication. It means that the organisms that best *fit* into their environment will *survive* to reproduce and pass on whatever features allowed them to fit better in the first place. In other words, if you happen to have some attribute that allows you to find more food, avoid predators, or get more mates than other members of your species, then nature probably won't seem as bad to you. And in all likelihood, you will have more sex and make more babies, and these babies will be like you and find life slightly easier than other babies. They will then give you more grandchildren and great-grandchildren, and on and on the process goes until one day, hundreds of thousands of years later, someone invents language and says you lounged about in a garden with a leaf over your genitals.

But what happens when the garden changes? We know the Earth isn't the same as it was when it first began about 4.5 billion years ago; in fact, the Earth's climate has bounced around between being a giant snowball and a swampy greenhouse for its entire history. There exists a reassuring turn of phrase, this one coined by H. G. Wells, which life uses for guidance during these times of great change—"Adapt or perish." Sorry, did we say that phrase was reassuring? We meant *unsettling*.

This expression isn't really advice for an individual trying to cope with rapid climate change, although it could be... It is more a description of the two possible outcomes for a species when the equilibrium of nature is disturbed. Go back a couple of paragraphs and think about all those gifts that allowed you to make more grandchildren because nature was kinder to you. Then imagine your environment burns down and suddenly those gifts are now curses. All those babies you made are now prime targets for predators, and you go from being a prehistoric Ghengis Khan–level reproducer to essentially a castrati. Burrowing Billy over there, once looked down on as a dirt-encrusted loner, is now sitting in the driver's seat when it comes to making the babies for the species, and over time, we all become mole people instead. But hey, at least you didn't perish like the other 99 percent of species that once called Earth home.

Let's just let that sink in. The about ten million plant and animal species alive today represent only about 1 percent of the plant and animal species that have ever lived on Earth. That means there is a 99 percent fatality rate for any new species, which probably extends out to 100 percent given a long enough time period when you consider how often we have to adapt or perish. It is doubtful that even the most hopeless gambling addict would risk the lives of all his children and grandchildren on a less than 1 percent chance of a payout, but the Universe forces us to play these odds every time we bring a new life into this biological arena of death.

Nature is difficult to survive in at the best of times. Even if you are lucky enough to be dealt a kick-ass hand with all the best cards, nature flips it on you, and without notice you are suddenly playing Twister, and those cards you were holding are making it hard to put your right hand on green. But if it makes you feel better, know that every other life-form is in the same boat. Except that boat is overturned and we are all treading water using only our legs. But remember, sometimes you just need to climb up onto one of the other suckers and let them do the work for you. It's what the Universe wants.

REASON

Everything will eventually suffocate

I bet you thought this chapter was going to be about how humans are affecting Earth's atmosphere—how very human of you. But the story of how our current atmosphere came to be is way more awesome. The formation of this thin, vulnerable, yet dynamic layer of gas floating around us—despite the repeated efforts to rid us of it—is a great example of the Universe's hatred for all things metabolic.

Our Earth hasn't always existed in its current form. In fact, around 4.5 billion years ago, there was a precursor to the Earth, which scientists refer to as the *proto-Earth*. This object formed shortly after the very beginning of our solar system, with some estimates suggesting that its formation could've taken as little as five million years. The continual bombardment from remnants of the formative solar system left the proto-Earth with a *steam atmosphere*, which in turn trapped the heat. The embryonic planet separated into layers as it started to cool, with the heavier, denser elements starting to sink toward the center as the steam turned to liquid water. This may give you the impression that it was like a Norwegian bathhouse, but take one of the coals from the sauna and shove it up your ass, and you would still be more comfortable than if you were walking on the surface of the proto-Earth.

Scientists believe that the most likely scenario is that at some point, an object the size of Mars smashed into our planet. Earth blew open like a Ford Pinto being hit by a moped in an event referred to as *the Giant Impact Hypothesis*—and you know when they use the word *giant*, they mean fucking huge. It is believed this collision and the scattered debris resulted in the formation of our current Earth-Moon system.

The rogue moped planet goes by the name *Theia*, which is based on Greek mythology. Theia is the mother of Selene, the goddess of the Moon—hence the rogue planet's name. So poetic! Anyway, Theia crashed into the proto-Earth, most likely texting while she was drunk on some cosmic alcoholic gas cloud. Not that we mean to be preachy about texting and driving, but Theia's now dead. Lucky for us, the collision led to the formation of the Moon, which has been by our side ever since.

As Earth's structure came to settle again, the relentless bombardment of asteroids and volcanic activity would continue for a little longer. This wasn't such a bad thing, as the volcanic activity would deliver the heavier gasses necessary for early life to form. Around 3.7 billion years ago, it is believed that life appeared deep within Earth's oceans. Interestingly, the first organisms to use light to get energy weren't plants; they were bacteria. And instead of producing oxygen, they made a salty acid. If that sounds unpleasant, we recommend you don't go looking up how anything fermented is made.

Fast-forward another billion years, and the stage was set for the construction of an atmosphere more conducive to the life we see today. *Cyanobacteria*—organisms that are still around today—started populating beaches all over the world. These little light gobblers used the Sun's energy to fill our atmosphere with oxygen. Hooray! Well...for modern-day organisms. There was no "hooraying" for the microbes who found the increasing amount of oxygen toxic. For them, it was the beginning of a long and worsening uphill struggle.

The oxygen started making its way to the upper atmosphere and formed the ozone layer, something that to this day protects life from the Sun's harmful rays. This layer of gas would also set the stage wherein humanity could demonstrate their stupidity on a planetary scale by wearing too much hairspray in the '80s. Propellants in aerosol cans, refrigerators, air-conditioning units, and other wonderful inventions put chemicals called chlorofluorocarbons (better known as CFCs) into the atmosphere. While they are harmless and nontoxic to humans, they apparently fucking hate ozone and blew a hole in our lovely blanket. Luckily, we went about addressing this issue, and the hole is beginning to close up again. Had we not acted on eliminating CFCs, we could have created an entire Australia-like planet wherein people walk around in flip-flops and beer-soaked singlet tops saying things like "Crikey! She's a bloody scorcher today, ain't it mate?" While the prospect of crocodile wrestling being added to the Olympics sounds alluring, be thankful we avoided creating this particular hell on Earth.

In an illustration of the temperamental nature of our atmosphere, around 2.5 billion years ago, the Earth then became encased in ice. This was due to the increase in oxygen levels and the subsequent decline of the greenhouse gasses keeping the Earth warm—say hello to *Snowball Earth*.

Interestingly, greenhouse gasses being pumped into Earth's atmosphere were a catalyst for early life to repeatedly flourish. Unlike today, these emissions that warmed the Earth would not be from "environmentally conscious" fossil fuel industries but from ancient volcanoes. The glaciation cycle happened a few more times in the past billion years, freezing over the Earth and pushing life to its limits. To make sure this doesn't happen again, humanity, along with the tireless work of lobbyists, is hell-bent on destroying every last glacier. Checkmate, Universe.

About 700 million years ago, oxygen levels in the atmosphere and oceans started to significantly increase, reaching about a fifth of what they currently are today. This increase created a huge divide in the life-forms that would dominate the Earth—those who could use oxygen to create energy, and those for which oxygen was poisonous. Luckily for us, Team Oxygen won out. But unfortunately, that also meant we were stuck with one of the most volatile chemicals in the Universe. (That's right, oxygen is a ruthless killer—see Reason 2.)

This massive uptick in atmospheric oxygen fueled a biological big bang known as the *Cambrian explosion*, which happens also to be the name of a local band, craft beer, barbershop/espresso bar, an avocado toast option from the café at the ax-throwing club, and some other stuff millennials are ruining. This occurred about 540 million years ago and is the origin for most of the major groups of animals that first appeared in the fossil record. If there was a creator, clearly they had been taking some strong hallucinogens during this time since some bizarre animals emerged during this period, including a five-eyed crab-worm with tiny clawed hands on the end of a trunk, which resembled a four-year-old's attempt at a drawing of a giraffe after watching too many Tim Burton movies. Another creature was so trippy scientists straight up named it *Hallucigenia*. But, from that point on, our atmosphere appears to have settled down. As it currently stands, our atmosphere today consists of approximately 78 percent nitrogen, 21 percent oxygen, 0.9 percent argon, and 0.1 percent other gasses...not to mention dust, pollen, tiny fecal particles, and probably some coronavirus.

All these changes in our atmosphere, though they led to the biodiversity we see today, illustrate just how *inhospitable* the *most hospitable* place is for us in the Universe. What a species we are... We look at all this chaos and helpless fragility, but only see it for its beauty. We are so far just a blip in the story of the Universe, and for all we know, it could end tomorrow. There are countless instances in the cosmos where a cosmic ray jet, solar flare, or gamma ray burst has rid a planet of its atmosphere. If you are playing the numbers, it's going to happen to us at some point, but betting on extinction is something even gambling addicts avoid. Until then, we better desperately hold on to this blanket that gives us warmth and protection from the harshness of the Universe, even if it is made mostly of poison.

PART V

BUT THERE IS NOWHERE ELSE TO GO

Finally, the space stuff, right?

REASON

Mars is an unlivable rocky shithole

Mars has always captured our imagination. The earliest film about Mars, creatively called *A Trip to Mars*, was about a professor who mixes some stuff together and reverses gravity. He then ends up on Mars, and like a bad acid trip, he loses his shit, runs through a forest of semi-human monster trees, finds some rocks that talk, and sees gigantic aliens... Hang on... This movie sounds epic—be right back... Never mind, it was terrible. In the end, after all the weirdness, he ends up back in his lab all dizzy. All of this makes you wonder what filmmakers were on in 1910. Whatever it was, I bet it's illegal now...lucky bastards. Unfortunately, the Universe won't provide us with a Mars of such awesomeness. Instead it's just a barren wasteland of a planet full of rocks, ice, extreme temperatures, and a shitload of radiation—essentially what we're trying our best to turn the Earth into.

Despite our more recent understanding of Mars's inhospitableness, our infatuation with the planet has never seemed to waver. We have written countless novels and made numerous movies about humans venturing to the red planet. It is safe to say that the moment we walked on the Moon, we focused our dreams on Mars. Even though we have not yet landed humans on Mars, our love for this mysterious planet is still very apparent. In fact, NASA has sent over twenty spacecraft missions, including five rovers, to Mars. It is rather extraordinary that a life-form on one planet has populated another with robots—despite the hurdles set out by the Universe.

Getting computers to the red planet hasn't always been sunshine and rainbows, which on Mars is extreme UV radiation and coronal mass ejections. We've tried and failed spectacularly when sending some stuff to our neighboring planet. Take, for example, the Mars Climate Orbiter, which was launched in 1998 with a mission to study Mars from orbit and serve as a communications relay to Deep Space probes and the Mars Polar Lander, which sounds very ambitious. So what happened? Well...we kind of lost the probe.

You're probably wondering the same thing everyone else was—how the hell do you lose a $200 million science experiment? Well, there was a slight, shall we say, *problem* with its navigation. The British engineers working on the craft didn't convert their English units to metric, which, surprisingly, despite being American, is what is used by NASA. In short, this unit conversion failure is quite probably the greatest embarrassment in the history of space exploration. So now can we all just agree on how to fucking measure something? Actually, that's not fair—*we*, as in most of the world, have agreed on the metric system. It's only Liberia, Myanmar, and some other place that refuses to move past living in the dark ages of pounds and yard sticks.

Anyway, it is believed that the Mars Climate Orbiter didn't do what its name suggested and orbit Mars, but instead burned up in the Martian atmosphere, and with it, $200 million and the hopes and dreams of thousands of soon-to-be-unemployed nerds. But just like that creepy guy on the dating app who can't take a hint, neither the radio silence nor the outlying cost would stop us from fantasizing about a future with the red planet—raising crops together, making babies, recycling your own feces to use as fertilizer... It's clearly our destiny.

After sending probe after probe, and many late nights watching Mars from afar, we finally got a message back saying, *Come over, I'm wet.* That's right, our efforts appeared to have paid off when scientists found the first evidence of flowing liquid water on Mars—a major discovery and massive leap forward in our dream of sending humans to the red planet. The discovery of water in its liquid form captured the imaginations of intrepid scientists and psychotic tech billionaires the world over. Then, 2020's answer to a cyborg Bond villain—Elon Musk—said that within ten years he plans to get humans on Mars. Is that realistic? It all depends on how reckless we are, and we humans can be pretty fucking reckless.

Sending humans to Mars is far more difficult than just packing a bunch of astronauts into a rocket and sending them on their way. It would require us to overcome challenges never faced before, as the mission would be far more complicated than just using a simple aim-and-shoot method. For one, engineers and scientists would have to account for the lining up of both planets' orbital paths to ensure the safest, fastest, and most affordable route is taken.

Let's imagine for a moment that humanity not only had the technology, but also the political/billionaire willpower to send humans to Mars. A one-way trip to Mars would take about nine months. We could possibly get you there sooner, but this would require too much fuel. You would also be forced to stay there for a bit as Earth continues on its path around the Sun. This means that even when you arrive, you would have to wait about another three months before taking off again. All in all, a round-trip package to Mars—with accommodation and sightseeing included—would take *at least* twenty-one months.

Got the time? Great. We have liftoff!

Well done. You and your crew have overcome the nausea and relentless jolting of your spacecraft, and managed to make it into space despite the hundreds of parts of equipment that are statistically likely to fail with any rocket launch. So crack open the bubbly and enjoy the rest of your ride... Well, not really.

Harmful solar radiation continually bombards the capsule, and you and your astronaut buddies need appropriate shielding from the Sun's damaging rays. Luckily for you, the team has lugged some protective materials on board. But they could be a little hard to find among all the other bits and bobs sent with you to build the required infrastructure, equipment, and other essentials for a crew of six astronauts.

Fortunately, your playboy billionaire boss has ensured that your crew got everything needed for a successful Mars mission, and has sent sixty separate rockets. That's right, sixty rockets—each carrying a payload of about fifty thousand pounds, or the equivalent of launching a total of eight blue whales into space and then sending them on to Mars. That should be enough supplies for you and your crew to set up a base on Mars that others could visit after your journey there.

So here we go. You're now at the stage of the mission where you are entering Mars's atmosphere, or as it is lovingly referred to *"the seven minutes of terror."* Unfortunately for you and your crew, it is similar to what one would imagine "seven minutes in heaven" to be like if it were played at Rob Zombie's house.

Locked inside a metal container, you're hurtling toward the Martian surface all while the Universe is outside with a flame-thrower trying its best to get in and destroy you and everything else. Oh, and don't bother calling for help; you are too far away from Earth now to call for immediate advice. This also means that you are too far away for them to provide any real-time assistance. You are on your own for the next seven minutes.

Surprisingly, you've made it to the Martian surface. So off you go to explore this mysterious world... But wait. First, you need to adapt to the lower gravitational force. Mars has a tenth the mass of Earth and is less dense—as such, it has only about 40 percent the surface gravity of our planet. The lesser gravity results in lower bone density and reduced muscle mass as your body now weighs the equivalent of a ten-year-old girl's; this means you had better have packed the Bowflex, and you'd better use it two hours a day if you intend to return to Earth without your legs breaking beneath you.

The lower amount of gravity on the Martian surface and its weak magnetic field mean that Mars has a hard time holding onto its atmosphere. This has resulted in the Martian atmosphere having a volume less than 1 percent that of Earth, which is a big problem for the fools that traveled there. Living on Mars results in you and your crew being bombarded with harmful ultraviolet radiation, along with severe and sudden temperature changes that quickly turn the weather from a nice summer's day to an Arctic winter in a matter of hours. But it's not all bad. Once you get over the unbearable sudden coldness and skin cancers, you get to see some lovely auroras. Actually, it is all bad. The light show you're witnessing is dangerous solar emission interacting with what little is left of Mars's atmosphere, with the rest pounding you and everything around you harder than John McEnroe's racket slams into the grass at Wimbledon.

If all that wasn't enough to dissuade you, Mars's atmosphere is primarily made up of carbon dioxide—about 96 percent, to be more precise. This is a much higher concentration than what we have here on Earth—a mere 0.04 percent. But that doesn't bother you; you want to feel those beautiful cosmic rays through your skull, the sandstorm kissing your cheeks, and the carbon dioxide flowing through your hair. So off comes your helmet, and for that moment, you are one with the red planet.

Fifteen seconds later...

You collapse on the ground, screaming in agony, as your organs rupture from the intense change in pressure. Congratulations, you have now surpassed the Mars Climate Orbiter as the greatest embarrassment in the history of space exploration.

Space travel is out of the question

If you flipped right to this chapter, interesting choice—but you probably didn't, so you should've realized by now that Earth is a violent place, the Sun will kill us one way or another, and we've been doomed from the start. But there is a glimmer of hope. If we could leave the confines of the Earth and venture to other planets, we could start new colonies. Hopefully it goes better than last time. But to find those worlds, we have to be prepared to spend a lot of time in the void of space. A *lot* of time.

If our own technology isn't used against us by the Universe (which it definitely will be—see Reason 7), then one day it'll be good enough to get through vast, almost empty space. And if we somehow work out the kinks associated with living on Mars (we probably won't—see Reason 26), then we could hopscotch out of this solar system, planet by planet, moon by moon. But then what? Well, assuming we've worked out how to travel at the upper speed limit of the Universe, it should take us around four and a half years to get to our *nearest* star. But come on, stay hopeful.

In zero gravity, your blood pressure drops, so you can leave the Lotrel at home, and you don't have to walk anywhere because you can fly like Superman in slow motion. Sounds pretty sweet, right? But wait. In these conditions, your body stops making as much blood, which you kind of need, and you lose all your muscle mass from lack of use. This is why astronauts returning from the International Space Station (ISS) are always picked up in wheelchairs—they are physically too weak to walk and have low volumes of blood. But we're making it sound like space is a vampire sucking the blood and life force from people's bodies. No, not at all. Space is so much worse than that. Losing that gym body you worked so hard for is the least of your worries should you find yourself outside of that cozy spaceship, and if you thought sex was awkward enough in Earth's gravity, you should try it in zero gravity.

Stepping out into space without a space suit might be one of the most rock-and-roll ways to go. You've got fifteen seconds before you pass out from a lack of oxygen to the brain. If you think you're clever and hold your breath, your lungs tear open and bleed everywhere inside you. You've then got about a minute before death ensues. After this you freeze and drift around the Universe as a person Popsicle forever. Or until someone finds you. Or until you get vaporized by a star. Hail Sagan.

If that rock-and-roll lifestyle isn't for you, then you'll need a space suit. Your space suit needs to have radiation protection, impact protection from space dust, oxygen supply, and insulation so you don't freeze, but not so much that you cook yourself. Any incomings like water or food and outgoings like piss and shit need to be accounted for, depending on how long your spacewalk is going to be. Clearly we are in the future, so perhaps long space hikes will be a thing. It could be romantic—a low-gravity expedition with a loved one across a distant moon, listening to the sound of them shitting in their space suit through your earpiece.

But let's jump back on board the spaceship, because that's realistically where you'll be spending most of your time. Now, let's go have a shower to wash that space suit splash back off... except that you can't. Water is very scarce in space and very heavy to carry, so they can't bring the 450 milk cartons of water you need for your thirty-minute wet-down. No more showers or baths in space; you can just use standard-issue waterless soap and a towel, or maybe it's sponge baths with a vacuum hose from now on. How luxurious. Besides that, how the hell is water going to fall in a shower with no gravity?

In your ear there are three semicircular canals which are about as big as two-thirds the diameter of a small dog's eyeball. Inside these canals is a constantly moving fluid that tells our brain which way is up. It's how we remain standing and walk around. Now imagine if those organs just stopped working. Anyone who has been on a roller coaster knows what it's like. Floating around in space is like that *all the time*. Imagine feeling like you are perpetually falling, I mean you'd get used to it, but who would want to? It would permanently ruin the fun of roller coasters. Luckily Arthur C. Clarke—better known as Stanley Kubrick—figured out long before 2001 that building giant spinning rings would allow us to create artificial gravity.

Their fictitious designs (albeit both plausible and very impractical) mimic the effects of gravity by imagining spaceships that rotate as they move through space. This allows the passengers and crew to walk around inside the periphery of any interstellar vessel as if they were walking on the surface of the Earth. Except that this periphery must be exactly the same distance from the center of rotation and spin at a constant speed. Too close or too fast, and the artificial gravity will add weight to your body faster than eating a supersized quadruple Big Mac meal with a melted butter chaser; too far or too slow, and you will drop weight faster than the inevitable diarrhea you would get from eating such a meal.

So you've ended the space odyssey and arrived at the nearest star. Fortunately, humans have been colonizing the region for years and there are a bunch of *Bishop rings* to choose a new home from. Bishop rings are theoretical space habitats that resemble a massive truck tire, with a circumference about twice the distance a Proclaimer is willing to walk to be the man to fall down at your door. You choose one with a nice temperate climate that's a little more populous because you're excited to meet other spacefarers. But wait, you've read lots of science fiction; what kind of Spacepox are your new neighbors carrying? Or maybe it's *you* that brought Enterprise Ebola and will kill a bunch of other colonists. Not the best start to your new life away from the doomed Earth.

Astronauts have been bringing bacteria to and from the International Space Station for decades. Scientists have looked at a couple of species found on the ISS and concluded they had unique adaptations specific to the conditions of space. The ISS zips around the Earth about sixteen times a day at a distance about thirty times that of mile-high club members. So what about all those space germs that would have decades or centuries or even millennia to adapt to living in the Bishop rings? Who knows what nasty diseases could spring up? Hell, we have trouble keeping track of the flu every year right here on Earth.

Let's recap: To get away from the doom and gloom of our solar system and reach out to find a utopia in the galaxy, we would first need to develop the technology. No hurdles there. Next we would have to manage the psychological and physiological consequences of traveling through the void of space for an extremely long period of time. Tough, but doable. During the immense time spent hopping around the galaxy, we would have to be lucky enough to avoid the solar storms, space debris, and radiation. Crossing our fingers couldn't hurt right about now. Then we would have to find habitable locations or at least find the resources to build some ourselves. There's got to be some out there, right? Then our immune systems would have to be able to fend off cosmic chlamydia as well as other interstellar illnesses. Right, so who's in?

REASON

The Sun is an angry dragon that breathes plasma

Anyone who has ever experienced a prolonged power outage knows how frustrating it can be. If you don't have a gas stove, you're doomed to eat room temperature snacks and takeout. The food in your fridge will eventually go bad, you won't be able to heat or cool your home, you won't be able to have a warm shower, and, eventually, you'll die. Yep, power outages kill people. In February 2021 a storm in Texas resulted in an outage for millions of homes, which resulted in hundreds of deaths. While skipping the warm showers didn't directly kill them, not being able to heat their homes meant most died from frostbite and hypothermia, as well as other complications associated with our dependence on electricity for basic needs. Indeed, more people often die from a resulting power outage following a natural disaster than from the natural disaster itself. Now, imagine if the whole world suffered a blackout at the same time. You can't just drive to Aunt Maeve's house in the suburbs this time to wash your dirty gym socks (see Reason 6). But it's still a good idea to go and check on her, as she might be dead.

Global blackouts are a real possibility thanks to the fact that the Sun is really a massive dragon. Well, more accurately it is a swirling soup of superheated hydrogen and helium that creates tangled up magnetic fields that periodically snap, throwing tsunamis of electromagnetic energy at us. But to describe it as a colossal fire-breathing dragon is more interesting and sounds way more badass. Every eleven years or so, like an angry Smaug, the Sun becomes really active and spits dragon fire all over the solar system. In this fantasy, we are the dwarves, and thankfully these heated expulsions are pretty unlikely to hit little ol' us, unless of course they are aimed directly at us, which dragons tend to do.

The immense age of the Universe has turned all the stars into sadists, and even our local Charizard hits Earth with a destructive solar storm every twenty-five years or so. That's way more frequently than you thought it was going to be, wasn't it? The reason our history books aren't littered with tales of the Sun getting all Targaryen on our asses is because solar storms in the pre-industrial era went by relatively unnoticed. Energy from even the most severe solar storms is largely absorbed by our atmosphere, and any that reaches the surface does not interact with living things. So in the past, even when King Ghidorah at the center of our solar system sent his fire our way, historical accounts just describe the sky being pretty for a few hours. No death. No destruction.

But what if you're not on the ground? Our ancestors doodling on cave walls didn't have airplanes, let alone astronauts floating up in space. What happens if you were flying to New Zealand to get a real Middle Earth experience, and you were unlucky enough to be in an economy class toilet and get hit with a solar flare? At that height, you would experience about as much radiation in a few hours as is recommended for you in six months. If you think that would present obvious problems, you'd be right (see Reason 11 for more fun facts about death by radiation). But even then, the problems this could cause for us from a biological perspective pale in comparison to the problems it could cause from a technological perspective. That potentially mutagenic blast could very quickly turn your transatlantic flight into the second worst Carnival Cruise of your life.

Before we started boiling giant kettles of water to generate electricity, solar flares just weren't a problem. Even now, most that hit Earth are relatively small and localized, and we often have enough warning that they are coming that we can simply turn off our gadgets until the storm passes. Then there are the really big flares. These bad boys can reverberate around the globe, which would almost certainly lead to widespread failure of electricity systems. People standing close enough to shoddily wired high-voltage systems would be electrocuted, which is not ideal. Global telecommunication systems, which connect you to the Internet by the way, would just stop, cease, end, kaput, that would be it...but the effectiveness of the customer service department from your provider would remain about the same.

Assuming you're a fucking idiot, you probably don't realize that your food has to go from farm to processing plant, to distributors, to the warehouse, to the supermarket, with various transport pathways in between. The fallout from a massive solar flare would stop communication along these supply chains, and people would starve. The damage to industry and infrastructure could put human civilization back centuries, starting a new dark age. If you lived in a cold part of the world, you would have to start burning your shit to keep warm. This is quite literal, as all water pressure would be lost, so no flushing toilets, showers, or sinks, or drinking water for billions of people; the shit would pile up, but think about it, a cistern would make a pretty good fire pit. At least the sky would be beautiful for a few hours, but damn, you couldn't even post about it on your socials.

Such an event occurred in 1859. It is called the Carrington Event because some British guy, presumably named Carrington, saw it happen. During the event, auroras were seen across the entire planet, but at this time the only electronic devices were telegraphs. Numerous fires broke out at telegraph stations, and many people with telegraphs in their homes received an electric shock from their devices. When you think about people watching the light show in the sky, then rushing to send the telegram DEAR PATSY STOP ARE YOU SEEING THIS SHIT? STOP and instead getting shocked, it's pretty funny. However, scale that up to the number of appliances we have on or around a person today, and it's not so humorous. Maybe move that phone to the back pocket. The chances of getting hit by one of these in the next fifty years is as high as 50–50. With all the astronomically rare shit that could happen, this feels like a certainty.

There is yet another solar event that is at a level above a Carrington-sized event, but we've never seen one come out of our Sun...so far. *Superflares* are the biggest kind of solar flare there is. Some scientists suggest our Deathwing in the sky could have a superflare event every five thousand years or so, and if it is well directed, it could do everything described above and more, stripping back the ozone layer and decimating multiple ecosystems, especially those in the oceans.

The wonders of the industrial age, especially electricity, have brought a standard of living unimaginable only a few hundred years ago. But on a whim, should the Universe decide to tickle the nose of our local Draco, it might all be all sneezed away. The Universe watched as we built the modern world using the might of the electromagnetic force, allowing our entire civilization to be completely dependent on it. Now it is laughing at how hopelessly exposed we are. If this were a fairy tale, we'd be the young princess forced to live in a tower under the subjugation of a dragon. Unlike a fairy tale, though, there is no knight coming to save us, and the Wi-Fi could go down at any time.

REASON

We are all alone in the cosmos

If you have not already realized by now, the Universe is one coldhearted bastard about −454.76°F, or −270.4°C, to be exact. You would be forgiven for disagreeing with that statement when looking around at the diversity of life on Earth, or as the father of evolution Charles Darwin put it, "endless forms most beautiful."

The Earth is full of a variety of different life-forms, with many that are still yet to be discovered—though we may be killing them faster than we can find them. Our current best estimates are that there are between 5.3 million and...wait for it...1 trillion species of life on Earth. That is a pretty big margin of error. It's not because scientists are bad at their job—though some definitely are—it is because there are still vast sections of the Earth that remain unmapped, unobserved, and unexplored. For example, we have only observed and mapped 20 percent of the world's oceans. One can only imagine the diversity of life that we have yet to discover lurking in the depths—plus old boots and shit tons of plastic.

Alternatively, if we look out into the cosmos, we see that life appears to be an extreme rarity. So far our best measurements and surveys have not detected life outside of Earth, and it's not for a lack of trying. The easiest way to think about what specific signs of life might look like is to put yourself in the shoes of an alien that is searching for us—that is, assuming the aliens have feet. Or perhaps we are barking up the wrong tree, and we are merely living in a simulation (which may not be as fun as your Sims make it out to be—see Reason 39).

First of all you would be searching for water. Water is abundant in our galaxy; it has been detected on planets, moons, and interstellar dust. The problem arises when we look for it in its most conducive form for life—liquid. Our observations indicate that our planet is somewhat of an anomaly in that more than 70 percent of the planet's surface is covered in liquid water. But let's not get too excited about being special. Scientists have detected plumes of water being ejected like huge geysers from multiple moons orbiting other planets in our solar system. One of the best candidates for life in our solar system is Jupiter's moon Europa. Just under its icy surface, Europa may contain more water than even Earth despite being slightly smaller than the Earth's moon. Scientists at NASA are very interested in careful scientific examination of Europa to determine if there are any other signs of life.

The usual signs of basic life are the specific concentrations of methane, carbon dioxide, and oxygen, along with some other gasses that are detectable in the atmosphere. These gasses are called biosignatures and provide evidence that there may be life-forms eating, breathing, and farting on whatever world we are staring at. Knowing the concentrations of these gasses also gives us an idea of how abundant and active these life-forms are. Take humans, for example: we take in oxygen and breathe out carbon dioxide—as well as lies and ill-informed opinions, but let's not go there. However, trees take in carbon dioxide and store the carbon to help it grow while releasing the oxygen back into the air, which in turn we breathe in. By comparing the concentrations of chemicals in an alien atmosphere, we can determine if they are consumed or excreted into the atmosphere by some sort of extraterrestrial organism.

It is pretty safe to assume that there are no technologically advanced life-forms outside of our planet despite what some rum-soaked redneck or ex-Blink-182 vocalist tells you. At least, there doesn't seem to be any near us. Though, to be fair, looking around *on this planet*, some might question if there is intelligent life in our solar system at all. A *technosignature* is similar to a biosignature but differs in that if detected, it would provide evidence of not just life, but intelligent life in the cosmos—it would also make an awesome name for a Berlin nightclub.

Technosignatures consist of communication waves, such as radio waves, lasers, or even orbiting bodies like satellites. Observations of atmospheric compositions are also useful in detecting intelligent life. For example, large amounts of carbon dioxide with specific amounts of other materials could indicate evidence of industrialization. While carbon dioxide is naturally occurring, not all carbon is created equal. Carbon dioxide from fossil fuels contains carbon that is a little bit different, and scientists can detect this difference. If there is enough of this other carbon in

42 REASONS TO HATE THE UNIVERSE

the alien air, then it could imply that the alien planet has a problem with wealth disparity and the representation of climate data.

Unfortunately for the health of our planet, we are great at leaving these technosignatures in our atmosphere. If scientists could detect these signatures somewhere else in the Universe, it could give them a good indication of intelligent life—like humans who are slowly heating up our planet and choking ourselves out of existence. To date we have not come across any life-form this stupid...err...intelligent.

We have sent machines to a few other extraterrestrial objects to test their soil and air in the hope of finding evidence of life either past or present. These missions have sparked concerns over the potential contamination of these innocent planets, moons, and asteroids by humans. In fact, when we were making regular trips to the Moon, we left a lot of our shit behind. Oh, and to be clear, we are not using the term shit as in "stuff"—we mean literal shit. There are almost one hundred bags of human shit, piss, and vomit on the Moon. What a species we are!

Now imagine for a moment that there is intelligent life out there and they *do* have the technology to contact us. Better yet, imagine they can actually visit us. Why would they? We're disease ridden, war provoking, self-sabotaging, ecosystem collapsing organisms that poison our own planet for the benefit of only a few individuals. And what makes you think that they would want to visit *humans* anyway? Dogs and cats are far superior. We literally clean up their shit and give them food while they don't even pay rent, taxes, or vet bills. All they have to do is give us a little affection and we melt in their paws. Now, *there* is intelligent life.

Let's face it, any chance we had of making contact with

extraterrestrials was ruined in 1973 after the launches of NASA's Pioneer 10 and 11 missions. Both were sent to observe the regions of our solar system that were previously unexplored. Pioneer 10 would be the first human-made object to travel through the asteroid belt and gain up-close images of Jupiter and its moon Io, while Pioneer 11 made the first direct observation of Saturn. By 2003, signals from either spacecraft had become so weak that they could no longer be detected back here on Earth. Both Pioneer 10 and 11 are now traveling in interstellar space. Why is this a problem? Well, it is not. The problem relates to what Pioneer 10 and 11 spacecrafts had with them.

Imagine sitting at home minding your own business, as you probably are. You hear some beeping that is getting louder and louder. You look out the window and see a random box at the door. The box is blank and there is no delivery person in sight. Your curiosity gets the better of you and you open the blank package. In it you find a gold plaque. Seems harmless so far. Then you turn it over. It's at this point you realize what is on the plaque. You are unsure of what is more disturbing: the naked images of a man and woman, or the directions they left to their house. Perhaps you could argue the most unsettling thing about this whole situation is that someone spent the time engraving this into a sheet of metal. And this is essentially what Pioneer 10 and 11 were carrying—an unsolicited dick pic with directions to our home.

It may seem unsurprising to many people that our first real attempt to contact sentient beings from another world involved showing them our junk. It has been a popular method of introduction for generations of men. A more important question to consider is, if found, how will the aliens respond? They might return the favor and do the whole *you show me yours, and I'll show you mine* thing. Imagine that—a universe where all life-forms are just sending off nudes into the cosmos. The very first signs of extraterrestrial life could be an explicit picture. At least it would add an interesting element to the Mr. and Miss Universe competitions.

Maybe we humans came into being a little too late and missed the party? The earliest evidence for life on Earth dates back 3.5 billion years. Depending on your criteria, the earliest evidence of *Homo sapiens* is about 250,000 years ago—a mere blip on the cosmic timescale. With some pretty easy mathematics, you start to realize that our species has only been present for 0.00002 percent of the 14-billion-year life span of the Universe. Perhaps there were awesome galactic gatherings long before our solar system and Earth formed, and we simply missed them. And here we were hoping to be the party crashers!

Let's entertain the idea for a moment that the Universe does have some sort of consciousness and can control what goes on in the cosmos. What sort of asshole sticks every living thing on a tiny rock in the middle of nowhere with no one else around and then occasionally bombards it with asteroids, solar flares, and a whole bunch of other shit? We're not saying we don't deserve it—it's just... the Universe is somewhat comparable to that weird kid at school. You know. They were the one who would find enjoyment in burning insects and said creepy things about their cousin to the class and teacher. If a child displayed some of the behaviors displayed by the Universe, parents and teachers might go and get them tested before they end up on some sort of government watchlist. Then again, a person sending naked images of themselves with directions to their house would probably end up on that list too.

Black holes are firing jets of pure energy at us

Just when you thought the emissions from your uncle after his second pot of coffee and a large serving of chili eggs were the most powerful thing known to exist, along comes something astronomers call *cosmic ray jets*. To fully understand why your uncle's efforts—albeit still disgraceful—are not the most destructive thing in the Universe, we need to understand a celestial object known as a *quasar*.

First detected by astronomer Heber Curtis in 1918, quasars are extremely bright and thankfully distant objects with a mass around a billion times that of our Sun. At the heart of these quasars lives a supermassive black hole, and these are about as scary as they sound. Much like water circling before succumbing to the inherent void of a drain, so too do the gasses falling toward a black hole. One of the ways that large amounts of matter can be thrown into a black hole is when two galaxies collide—and if the reality of two galaxies smashing together doesn't give you some perspective on the cosmic insignificance of your Twitter spat with @FreedomFighter4913, then keep reading.

In what you could consider an ironic twist, if you were to search for a quasar (the brightest thing in the Universe), you would need to look to a black hole (the darkest objects in the Universe). As the black hole spins, gasses travel around it at different speeds dependent on their distance to the center. They bump and grind like enthusiastic extras in a Jay-Z music video, heating up and creating enormous amounts of energy. This causes them to glow, creating what is known as an *accretion disk* that can outshine the cumulative brightness of all the stars in a galaxy a hundred times over.

Like a debilitated tourist who ate a questionable sandwich from the airport café, quasars shoot large amounts of matter at high energy out opposite sides. And much like the tourist's decision to buy the sandwich, scientists have no idea why. These are the most destructive things in the Universe, and most people have no clue they even exist. To be fair, the plethora of channels showing beautiful people yelling at each other can sidetrack your journey to the Science Channel. But the next time you're binging some neurotic reality TV, we hope you reflect on how sociopathic actual reality can be. Interestingly, spinning black holes sucking everything around them into a reality that seems to defy all common sense, and then broadcasting it across the Universe, is a useful analogy for the phenomenon that is reality television.

Where these ejection sites are pointing will dictate how visible they are on Earth. If the quasar's cosmic ray jets are pointed directly at us, they get called *blazars,* which just edged out Quasar McQuasarface in the online naming contest. These cosmic ray jets consist of a variety of particles but overall are neutral in charge. The sheer magnitude of energy given off by the quasar—or blazar if it is pointed in your direction—is enough to accelerate the particles in these jets to around two-thirds the speed of light, which means they could make a full trip around the Earth in a fifth of a second. That sounds fast, but remember that porn travels around the Earth two and a half times faster.

If you were ever unlucky enough to have one of these things pointed right at you, it wouldn't be fun. First you would see the Sun's apparent brightness dwarfed by the incredible brightness of the blazar. The energy of the ray would be concentrated into something the width of only a few light days across, still large enough to fit the Sun and every planet in the solar system within it. As the jet hits the atmosphere, the gasses are excited to a point where they would no longer resemble a gas but rather a cosmic light show of death and destruction. Life would be absolutely disintegrated by the cosmic bombardment and the resulting Earth would have about as much activity as Lubbock, Texas.

At this point, you must be wondering how it took astronomers so long to detect something that we claim is worse than your uncle's efforts on the crapper. In all fairness to our scientific forefathers, the first direct image of a black hole was only taken as recently as 2017. When Heber Curtis first imaged the cosmic ray jets one hundred years prior, it was still a mystery as to what they were or what was causing them. Curiously, both of these observations, a century apart, were of the same galaxy, Messier 87, which is one of the most massive galaxies in our local group. At first, all that scientists knew was something very big and powerful was happening at the center of that galaxy. But after more than one hundred years of research, we now have a much clearer idea of where these jets come from.

Let's go back to the start. The infant years of the Universe, much like our own lives, were full of potential. Then came the obvious problem. The Universe hit its teens, and much like how a teenager's face explodes with pimples, the Universe exploded with quasars. It was super awkward, and even though the Internet didn't exist yet, the teen Universe was already beginning to show signs of its cruelty. If life had the opportunity to form during this period, it may not have lasted long.

Around four billion years in—when the Universe was probably getting asked to do something with its life by its parents—its acne (the quasars) started to clear up. Like the modern-day human, the Universe, having made it through puberty, was on to its next phase of its body transformation—continual expansion. As space expanded, so too did the distance that the light from the early Universe had to travel to reach us. There still exist quasars today and more will continue to form, but these will be rare, like an ingrown hair on the inner thigh and not the outbreak of pimples that took place in the early Universe.

Even though quasars are destructive, their raw energy can act like a supercharger for baby solar systems, helping them form stars, planets, and, in at least one case, the ingredients for life. As the Universe ages, the chances of being destroyed by quasars diminishes. It's past the volatile teenage period and has entered a period of calm, somewhere between the drunken early twenties and motorcycle-riding midlife crisis. We just happen to exist during this calm, low quasar-forming period. Maybe the Universe isn't so bad after all.

Ha! Just kidding—it's shit. If you didn't already know, allow us to break the news to you. Our precious Milky Way galaxy and its central black hole are on a collision course with the Andromeda galaxy and its black hole. And along for the ride is the Triangulum galaxy, but it's more like the keep-to-themselves cousin your mom forced you to bring on a road trip with your friends. The interactions of these galaxies could shift the stars, planets, and galactic dust around, causing one or both of the black holes to start consuming large amounts of matter again. Eventually these supermassive black holes will fall into each other and make an even bigger super-duper massive black hole forming a new quasar. But don't worry too much as this galactic collision is not scheduled for about another 6 billion years. The Sun, and probably we, are certain to be long dead by then.

REASON

All your favorite stars will die

Do you have a favorite star? There are so many to choose from. Perhaps your answer will be something obscure like Tommy Wiseau, or perhaps you'll go with a classic like Tommy Wiseau. The consensus over here is Acrux. It's the name of an actual star. Acrux is the southernmost point of light in the Southern Cross constellation. It has great cultural significance to most peoples of the Southern Hemisphere, appearing, for example, on the flags of no less than five nations. During Australian evenings, it serves as a guiding light to the nearest pub. Technically, Acrux is a six-star system, but let's ignore that fact. What's important is that Acrux, like all your favorite movie stars, will soon die.

As sad as that is, the reality is much worse. Although movie stars often go out with a bang, it is usually a metaphorical bang, like Paul Reubens showing his peewee in public. Acrux is going to go out with an actual bang, wherein we don't make a joke about someone getting shot on a movie set. The largest star of

Acrux is at least ten times the size of the Sun, which means it will die in a horribly cataclysmic event known as a *supernova*. Before you ask, no, our own Sun will not suffer the same fate—though you definitely don't want to be around when it ends the second half of its life (see Reason 33). Although you are often told, "*The Sun is just another star*," that seems to suggest that all stars are more or less alike. That's wrong. Who told you that?

Stars are so different that there are dozens of ways to classify them. The most common one you'll encounter is the *Harvard system*, developed by Annie Jump Cannon. The system categorizes stars based on their surface temperature. The hottest stars have surface temperatures up to 30,000 degrees Celsius, while the coolest stars are as low as 3,000 degrees Celsius. Our Sun is roughly 6,000 degrees Celsius. That sounds kind of simple, but there is an obvious problem. Like a movie star, how hot an actual star is changes over time. For example, Mickey Rourke's and George Clooney's hotness have both changed as their careers progressed. We'll leave it to you to decide in which directions. However, since the age of a star is irrelevant on the timescale of a human life span, let's pause and take a snapshot.

For about 90 percent of the stars we can see right now, the Harvard system tells most of the story. The hotter the star is, the larger and brighter it is. There is also *always* a simple relationship between temperature and color. Hot stars are blue. Slightly cooler stars are white. Moving lower in temperature, we see yellows, then oranges, and finally red-colored stars. Typically, the red ones are smaller. However, there are also monstrously large and bright but cold red beasts of stars, so who ordered that?

What happens throughout the life of typical stars is not that exciting—at least not if you are looking for an existential crisis to tweet about. As always with the Universe, there are exceptions to this rule, which unfortunately includes our closest star, which we call the Sun. It is going to take us for a pretty shitty ride in the future (and it's probably worse than you think—again, see Reason 33). But long before our own star begins to act up, there are other stars we really should worry about. These are the big fuckers, the supermassive stars. Betelgeuse, for example, is a red supergiant star in the constellation Orion. It is one of the largest and brightest stars in the sky, and if you say its name loudly three times, Michael Keaton will appear and lecture you on animal cruelty, which is a worse fate than anything we will describe here.

Betelgeuse is *only ten* million years old, and it's about to die. It started its life the same as any other star, as a ball of hydrogen gas. But Betelgeuse began with way more hydrogen—possibly twenty times as much as our Sun. You might expect that more hydrogen means more fuel to burn and, in turn, a longer life. In fact, the opposite is true. The gravitational force of all that mass means the core of baby Betelgeuse was much denser. The hydrogen fused into helium much faster, causing Betelgeuse to burn a hot bright blue. The core stopped fusing hydrogen a million years ago. But that obviously wasn't the end of the story.

The massive amount of energy produced by fusion inside a star pushes out against the massive amount of gravity pulling the star together in a sort of delicate balance. But after the core runs out of hydrogen, fusion ceases and gravity starts to win, crushing the core. This causes heavier helium to start to fuse while any leftover hydrogen on the outer layers puffs out and cools, turning the star red. From the outside, not much seems to change after that. But inside, it's a fusion party. The cycle of fuel exhaustion, core crushing, and new elements continues with shells of heavier and heavier elements being formed as we look deeper into the core. Eventually, the core will consist of iron. Iron fusion absorbs energy from the star, and so there is nothing left to counteract the force of gravity pulling the star in on itself. In a matter of seconds, the star collapses and most of its atoms rebound off the dense core and explode out in a spectacular blast we call a *supernova*—astronomy porn for us, but not so good for anything near the star.

We don't know what the interior of Betelgeuse looks like, but scientific modeling suggests it has no more than 1 percent of its life span left. Since Betelgeuse is about 650 light-years away, it may have already died, and we are just waiting for the explosion to reach us. That sounds bad, but just how much worse is a supernova than a surprise visit from Michael Keaton? At least in one of these situations, it will all be over relatively quickly.

At the moment of collapse, the core of a star can reach 100 billion degrees Celsius. That's thousands of times hotter than the core temperature of the Sun. With this amount of heat energy, subatomic particles called neutrinos are created and flung out in every direction. About 10 percent of the mass of the star is converted into this burst of neutrinos. Since neutrinos are tiny and barely interact with anything (sometimes they are referred to as *ghost particles*), many can escape and thus precede any other effects of the supernova. In fact, the SuperNova Early Warning System (SNEWS) is a global network of detectors that can give up to several minutes of warning if a nearby supernova occurs.

You can even sign up for the mailing list if you are keen on being able to predict the end of the world. You'd only have a few minutes, though—not really enough time to form an apocalyptic death cult or anything.

Following the neutrino pulse would be a shock wave of matter traveling near the speed of light, which would incinerate everything in its path. When you see a photo of a supernova, you are usually looking at the front of the shock wave, which can travel thousands of light-years. Luckily, you don't need to worry about the shock wave because the high-energy gamma radiation will get you first. One plausible candidate for the cause of the first mass extinction—which killed an estimated 60 percent of ocean life—was gamma radiation from a nearby supernova.

Estimates vary, but the most conservative distance beyond which Earth might be spared from a supernova is 1,000 light-years. Much closer than that and the gamma radiation would destroy half the ozone in our atmosphere and convert the nitrogen and oxygen into dangerous nitrous oxides, which are poisons (not really conducive to life—see Reason 16). There are six candidate stars that could threaten Earth, including our friend Betelgeuse. So, it's not like we are in the clear for another supernova extinction.

At 650 light-years away, Betelgeuse's neutrinos and gamma rays may leave their mark. It will certainly appear in the sky as bright as the Moon for several weeks. This of course means we will be able to see it during the day. That will be followed by the formation of cults and conspiracy theories, which sadly can have noticeable political and public health effects. And then, long after the brightness has faded and the event is forgotten, the shock wave of charged particles will finally arrive one hundred thousand years later. Luckily, that other ball of burning nuclear death, our true favorite star, will protect us with its own flux of solar energy...that's also probably killing us (sigh—see Reason 11).

REASON

Giant space rocks are constantly crashing into us

There was a day, about sixty-six million years ago, when you probably would not have wanted to be a dinosaur—except maybe a T. rex because they are fucking badass.

It is an objective fact that dinosaurs are pretty cool. Ask any five-year-old—they don't lie. These prehistoric creatures are also pretty impressive. The name *dinosaur* loosely translates to *terrible lizard*, and they are unquestionably the largest things to ever walk the Earth; the largest weighed about ten times as much as the biggest elephants and were three or four times taller than a giraffe. They ruled the Earth for at least 165 million years. Just pause for a moment and reread that—165 million years. Civilizations seem to only last several hundred years. We are pathetic next to the dinosaurs. [Cue Jurassic Park theme music.]

Yet we are here and they are not. So maybe we are doing something right—or something went horribly wrong for them. Not counting the current mass extinction humans are causing, the last mass extinction was the one that wiped out our dear feathered lizards. (Yeah, they have feathers now—Steven Spielberg lied to you.) So what caused their demise?

Let's dismiss a few candidates that might pop into your mind. First, comets. Comets, at least the ones that make the news, are relatively rare. A comet is a clump of ice and rock left over from when our solar system formed. They spend most of their lives in the far reaches of the solar system where the Sun's energy is too weak to disturb them. But many have orbits that bring them close to the Sun and among the planets. As they near the Sun, they begin to warm up and release gasses from their ice vaporizing. This gives comets their trademark "tails," which glow bright enough to be seen by the naked eye.

Since the brightest comets can be seen even during the day, humans have been witness to many over our existence. And if you've read any history, you are well aware of the imaginative things our ancestors took them for, but what can you expect from an age where our best theory for the Moon was some spheroid goddess who requires sacrifices for some reason? Indeed, every time Halley's Comet flies by—about once every seventy-five years—it is accompanied by disasters. Most are obviously coincidental (bad crops, too much rain, not enough rain, dead emperors—that sort of stuff), but some are causal, since stupid humans took it as yet another divine sign to kill themselves or other people, which, sadly, we are famous for.

It was actually Edmund Halley himself—for whom the comet is named—who showed that Halley's Comet had a parabolic orbit and predicted its period of seventy-five years. We'll do the math for you. It was last seen in 1986, so it will return again in 2061. If you are reading this in 2061, good on you. If any of the three authors aren't dead yet, we owe you a beer—or whatever the digital equivalent of beer is in the Matrix. Don't worry, though, Halley's Comet won't ever hit Earth. In fact, that any comet will hit Earth is pretty unlikely. The reason is simple: there just aren't very many of them, and Earth is a puny little rock floating in the vastness of space. Imagine trying to hit a moving grain of sand with a speck of dust from opposite ends of a football pitch. We're not saying it's impossible, but it's less likely than us buying you that 2061 beer we just promised. Astronomers have tabs on about a few thousand comets, so we know in the grand scheme of things, a collision is extremely rare. Besides, we'd definitely see it coming because a comet is basically a flying lighthouse.

The next things to rule out are meteors. This one is simple. Meteors are small pieces of rock that burn up when entering our atmosphere. Moving through the air causes friction, which heats things up. Don't worry, unless you are hurtling toward Earth in a spaceship, you'll never notice this friction for yourself. But meteors can travel over 40 miles *per second*. This turns them into—quite literally—balls of fire flying through the sky. Again, this would be cause for concern if you looked to the heavens for advice on how many children you needed to sacrifice for rain. But since meteors are small, they are not an existential threat—at least not to whole populations. A meteor *may* have killed *a* dinosaur, but definitely not *all* the dinosaurs. The meteors that actually make it through the atmosphere and hit the Earth are called *meteorites*. The biggest ones are about as large as a tank, certainly big enough to do some damage, but not enough to cause a global mass extinction. The dinosaurs lost last time, but maybe the next one will at least spare the noble *antechinus* (yeah...you remember Reason 17).

What you are looking for if you really want to ruin someone's day is an asteroid. *Asteroid* is a funny-sounding name for a massive rock that many people assume hurtles toward Earth from the outer reaches of our solar system. But most asteroids are actually not found around Uranus (those are hemorrhoids). In fact, most asteroids in our solar system are in the aptly named *asteroid belt*, a much closer ring of rocks found between the orbits of Mars and Jupiter.

Asteroids in the asteroid belt come in many sizes. The smallest ones are indistinguishable from meteors. These are tiny ones that burn up in the atmosphere quickly and appear as streaks in the sky—also known as shooting stars. The biggest object in the asteroid belt is Ceres, and it's an absolute beast of a rock. It was the first asteroid ever discovered and is about 1 percent of the mass of our Moon. It is about as wide as New Zealand and would probably take an entire tank of gas to drive across. It holds about a quarter of all the mass in the asteroid belt, and many astronomers consider it a dwarf planet, which is the same status Pluto has. If Ceres ever ran into Earth, we'd be truly fucked. Two planet-sized things smashing into each other is probably something you don't need to simulate to find out what survives...except, of course, if you are the Discovery Channel and need some terror-inducing stock footage to run before your documentary on ancient alien Egyptian megastructures.

In 2005, the Discovery Channel ran a simulation of an asteroid the size of Ceres hitting Earth. The video is a recurring viral

Internet phenomenon, which you can find many copies of on YouTube played against Pink Floyd songs, but the gist of it is that the entire surface of the Earth (which contains all life) would be disintegrated in minutes. But is this just standard dramatization? No. Geological evidence suggests Earth has been struck by asteroids this big several times. However, this was also billions of years ago when the Earth and solar system were still forming. Today, scientists are pretty certain such an event is never going to occur in the remaining lifetime of the Earth. The reason is again simple: we know of pretty much every object larger than 100 kilometers wide in our solar system. There aren't very many nearby, and none are on a collision course with Earth. Phew.

Not so fast though. The asteroid that killed off the dinosaurs was nowhere near that size. It was only about as big as a mountain, not an entire country. Of course, we don't know *exactly* how big it was. It hit Earth sixty-six million years ago, so no one was around to record it—nor would they want to have been. Also, the object is not lying at the bottom of the ocean or under the ground—it was completely obliterated on impact. What we can measure exactly is the 180 kilometers wide Chicxulub crater in Mexico that it created. Clearly, since we are talking about it, such an event does not immediately kill all life on Earth—so that's good—but it was still a shitshow.

When the asteroid struck the Earth, it disintegrated and liquefied the ground around it, instantly wiping out everything in an area about the size of Vermont. The shock wave raced around the globe, shredding everything in its path to pieces. The volume of stuff that existed where the crater now stands was blasted into the atmosphere and rained fire down on the entire globe, starting forest fires that essentially burned everything. Trailing behind to put the fires out was a 100-meter-high tsunami that drowned anything that wasn't on a mountain. And this was only day one. Over the next few days, the upper atmosphere was blanketed with dust and ash that blocked the Sun. Without sunlight or heat, the remaining land plants and most ocean plants died, destroying essential food for everything on the upper end of the food chain. In exactly 3,467 years, 2 months, 7 days, 3 hours, and 17 minutes after impact, 75 percent of all living species went extinct, and it took millions of years for life to recover. And the lack of T. rexes roaming around tells us it was never the same again.

If another 10-kilometer-wide asteroid struck Earth, humans would have a low chance of survival. So that's not ideal. Luckily, these extinction-causing asteroids only hit Earth once every hundred million years or so. More reassuring, NASA thinks it has tabs on 95 percent of all near-Earth asteroids—28,460 of them as of this writing—and there are only four that are bigger than 10 kilometers. If the Universe were trying to kill us *all* with asteroids, it's not doing a very good job of it. Having said that, in 2019 astronomers found an asteroid about the size of a football field. The problem was that they only found it after it had sailed past Earth at a distance described by many astronomers as "too close for comfort." Perhaps the Universe is still warming its arm up.

Maybe the Universe is just toying with us instead of trying to murder us all. The Earth is struck by about 100 tons of rock and dust every day, and most of it burns up in the atmosphere. It's estimated that hundreds of meteorites, if not more, make it to the surface of the Earth every year. One of these would be enough to fuck up your weekend. But this isn't all about you. There are more than two thousand near-Earth asteroids designated as "potentially hazardous objects" that would fuck up all our weekends. These asteroids need only be about 140 meters wide to destroy a city, which we hope is yours, not ours... unless we live in the same city, because none of us have asteroid insurance.

PART VI

THE UNIVERSE IS GOING TO WIN IN THE END

Changed our minds; space is scary as fuck.

REASON

The Sun will die and take us with it

It's amazing that we've worshiped this indifferent ball of exploding hydrogen for so long, only to find out it doesn't give a single shit about us. The truth of it is that the Sun won't last forever. It will eventually die, leaving the solar system a cold graveyard of forgotten dreams and lonely robots. If that's a shock to you, you are probably wondering if *future* means, like, *tomorrow*. We're not giving away the punchline just yet though.

Humor us for a moment while we present a situation that you will either find incredibly stupid or pedantically interesting. *What is the probability that the Sun will rise tomorrow?* Even the smartest people throughout history could not decide on the answer to this question. In fact this question is so famous, it even has its own name, the unimaginative *sunrise problem*. Every sane person's answer is 1—that is, they would say it's a certainty. Statisticians disagree. *"There is not enough data to be so certain!"* they say. Sure, you've seen the sunrise every day for the past few thousand days—and people before you have witnessed it rise for several thousands of *years*—but that only suggests the probability is high, say 0.9999999999, but not 1. Reason alone cannot tell you the Sun will rise tomorrow.

Allow us to elaborate further. Consider the following *white swan* metaphor (entirely meant to be confused with the *black swan* metaphor). For Australian swans born before the late nineteenth century, every swan ever seen was black. Should native Australian swans have then assigned a probability of 1 to the event that the next swan they would see is black? Clearly, with the benefit of hindsight, the answer is no. In 1896, British colonists brought white swans to Australia to steal the land and resources of the black swans. This part is not a metaphor, but also kind of is. So just because every swan you've ever seen was black, doesn't mean the next one couldn't be white, which—like the Sun not rising—is probably a bad thing for a land of only black swans.

The moral of that story is *people are shit* and swans are shit at assigning probabilities to things. In fact, the statisticians were right about the answer but wrong about why. The reason the answer is not 1 is that there will be a day in the future where the Sun will not rise. Will it be tomorrow? Well, that all depends on what day it is *today*. If you are reading this on the 9th of March in the year 5,500,002,022, we've got some bad news. Actually, that's a bit of a lie. If you are still here on the last day of Earth, you will see one last *epic* sunrise.

In roughly 5.5 billion years, the Sun will have expanded, engulfing Mercury and Venus, and most likely Earth. At that point, the Earth will fall into the core of the Sun and disintegrate.

Not even Paul Rudd, who will still look thirty-five at that time, will survive. If you end up making it that far with Paul, it would be pretty impressive. Not only would you have lived 5.5 billion years past your expiration date, but you'll also have survived the oceans being boiled off billions of years sooner.

Now, all this jumping around in time is starting to get confusing, so let's start at the beginning. Our solar system was formed about 4.5 billion years ago. Like all stars, the Sun's "birth" was marked by the first fusion of hydrogen into helium in its core. The fusion is ultimately powered by gravity, pulling all the mass of the Sun toward the center. But when two atoms fuse to create another element, energy is released, a lot of it. So much, in fact, that it pushes back out against the gravity that squeezed it together. These competing forces eventually equalize, resulting in a perfect glowing ball, what we would call a star. So long as hydrogen is available in the core, the Sun will continue fusing it into helium, radiating energy that brings light and warmth to its orbiting planets. Indeed, our lovely Sun has been the primary source of energy for life on Earth ever since the very first single-celled organism swallowed a sunbeam and farted out a copy of itself.

You may already see a very simple problem with this so-far happy story. The Sun is big but has a finite size. That means it has a finite amount of hydrogen. And if it is turning hydrogen into helium, then...yep, you got it—like a tire fire that's run out of tires, the fun eventually has to stop. Unlike a tire fire, though, as the Sun loses hydrogen, it actually heats up! As heavier elements that are not contributing to fusion build up, gravity pulls the core closer together, which forces what little remaining hydrogen exists in the Sun to fuse faster, which causes the whole star to heat and grow (if your dying star is shrinking, you've got bigger problems—see Reason 31).

So, long before anything spectacular happens, the Earth will simply get too hot. The polar ice caps will melt, the oceans will evaporate, and the atmosphere will thin as most of it is blown away. At this point, the Earth will be a desert, like the setting of a sick dystopian sci-fi movie starring Paul Rudd—he's so versatile. Perhaps carbon-based life will have figured out some way to survive on Earth, but it's not likely. Maybe our robot progeny will be around to witness the final throes of the Sun. What they will see is sure to be the greatest light show on Earth.

As the core of the Sun continues to collapse, the pressure and heat grow to the point where helium starts to fuse into carbon and oxygen. The core will become unstable and pulse with explosions of fusion, which blast the outer layers of the Sun out into the darkness of space. All the dust and gas will glow and be seen many years later from other parts of the galaxy as a beautiful stellar nebula signaling the murder of trillions upon trillions of life-forms, and the Universe still won't care.

The leftover core of the Sun, at this point only about the size of the Earth, will still faintly glow. It will be what any human left would call a white dwarf star, slowly cooling and dimming—an agonizingly long last breath in the death of an otherwise unremarkable star. Basically, the Universe has set us up for failure. The Sun is a Trojan horse. Don't trust it. Even Moses knew this when he wrote Genesis: *And God said, 'Let there be light,' and there was light. And God saw the light, and it was good. And God didn't know about fusion, and God fucked us all.*

A black hole could show up at any minute to rip us apart

Jaws was one of the most impressive movies of all time. Before *Jaws*, sharks were lovable creatures of the deep that occasionally mouth-massaged a surfer or two. But after *Jaws*, the world was deathly afraid of what lurked in the dark waters. Spoiler: it's not actually sharks. But if you are still afraid of dark waters, we have bad news. What lurks in the darkness of space makes Steven Spielberg's ani-matronic sharks seem like Furbies, either of which is the stuff of real existential nightmares.

The sharks of the Universe are *black holes*. They lurk in wait until the time is right. Then, without warning, they swallow a beach full of attractive movie extras. But what are black holes anyway? Simple. A black hole is a star that doesn't shine. It looks like literally nothing. It emits no light—hence the *black* part. The *hole* part can be chalked up to artistic license, so we're going to ignore it. So how can a star be black and emit no light? And for that matter, why is it so scary? We're building up the tension, don't worry. *Da dum, da dum, da dum...*

A black hole is what is left behind when a big star dies. Yes, all stars, including our precious Sun, will die. Actually, the result of a star dying is not always a black hole. Our Sun will become something infinitely more boring when it dies—a lowly white dwarf, a hot ball of carbon and oxygen that will spend the rest of eternity slowly cooling down with its feeble faint glow. If the star is less than ten times the current mass of our Sun, it will also end up as a white dwarf. If it is between ten and twenty-five times the mass of our Sun, it will end up as a *neutron star.*

A neutron star begins its life as a massive star, which is unimaginatively called a *supergiant* star. Being large, its immense gravity forces its own atoms to squeeze together and fuse into new elements very quickly. The largest stars, for example, may last *only* ten *million* years. That sounds like a lot, but remember that our Sun will live for ten *billion* years. Ten million years ago, the dinosaurs were still long dead and the last common ancestor of humans and most apes stood up to pee, a great evolutionary adaptation to be able to write your name in the snow. This means there are a lot of dead stars out there...and a lot of pee.

When a supergiant star finally runs out of fusion fuel, the result is a nova—but not just any nova, a *super*nova! The star collapses into itself and then rebounds in one of the most violent events in the Universe (not something you'd want to witness up close—see Reason 31). The neutron star left behind is only slightly larger in mass than that of our Sun, but it is all packed into something the size of Staten Island. This is incredibly dense. If all the living things on Earth were squashed together to the density of a neutron star, the resulting mass would be about as big as your average turd. Incidentally, this is about half as dense as your average vegan's stool.

Going back to the *Jaws* analogy though, being hit by a

neutron star is not like being attacked by a shark at all. It's more like being hit by the boat (the bigger one, of course). We would definitely see it coming, and it would just smash into us like, well, being hit by a supermassive boat. Unless you have an unnatural fear of boats, this is not the existential crisis you masochistically crave. You are here for the good stuff, like what happens when a star more than twenty-five times the size of the Sun goes pop.

A black hole begins its life the same way as a neutron star, a supergiant ball of hydrogen, but bigger. It burns through its fuel quickly, albeit a bit faster on account of its size. It explodes in a supernova in much the same way as a neutron star's progenitor. However, because it has so much mass, the effects of gravity become so extreme that they are unavoidable. In a certain region around the mass, nothing can escape the pull of its gravity. Not even light—the fastest thing in the Universe—can escape.

The boundary beyond which escape is impossible is called the *event horizon* because events inside of it can never be witnessed from the outside. If you have seen a picture or illustration of a black hole, it looks like a glowing black ball. The glowing part is light from the extremely hot stuff buzzing around just outside the event horizon. The photo you've probably seen (as it is the only one ever taken at the time of writing this) was from the Event Horizon Telescope of M87*, the black hole at the center of galaxy Messier 87. This galaxy is in the constellation Virgo, which makes it hardworking, funny, and a great lover. It also means that it is super-compatible with objects in the Taurus constellation, which are graceful, diligent, but stubborn at times. Just watch out when the Sun passes through Aquarius, which if you haven't rolled your eyes yet means you're a fucking idiot.

M87* is what is known as a supermassive black hole because it is several billions of times larger than the Sun. Its event horizon would extend many times the size of our entire solar system. These supermassive black holes are suspected to exist at the center of every galaxy, including our own Milky Way. They get to their size by swallowing other stars or merging with other black holes. They are the leviathans of the Universe.

But supermassive black holes are not the ones you need to worry about. First of all, they are huge and obvious. We'd definitely see one coming. Second, there aren't that many of them, relatively speaking, so they are quite thinly spread out across the Universe. Lastly, the death they would cause—and you would surely die if you approached one—is quite boring. You would simply be incinerated by the hot ionized gas swirling thousands of kilometers per second around it. You'd never even make it to the event horizon, which is where the fun begins.

Luckily, for the sadists anyway, there are billions of smaller black holes for every one supermassive black hole. For those keeping score, yep...there are billions of black holes in our galaxy right now. These are the sharks lurking in the depths. How often does something swim up to us? So far as we know, a rogue star—of any type—swims by every hundred thousand years or so...and we are probably due for another one soon. So what will happen if a black hole shows up? I'm sure you can guess it's nothing good.

The strong gravity of the black hole would extend beyond the event horizon, so shit would start hitting the fan quickly. Just as the measly Moon can pull the tides, the black hole would first vacuum up the atmosphere and oceans. The huge shift in water would look to the little creatures on its surface as massive tsunamis. At the same time, the part of Earth closest to the black hole would feel more gravity than the farthest. This would cause internal stress leading to earthquakes and volcanoes. Most likely everything would be dead at this point, except for those few billionaires who spent it all building bunkers to enjoy that self-righteous feeling of living a few minutes longer than everyone else. Don't worry if you aren't a billionaire though; they won't survive either.

Eventually the Earth itself would start to deform, becoming more ovoid as it got nearer and nearer to the event horizon. As we got closer to the center of the black hole, the pull of gravity would become more extreme, meaning things that were closer would get pulled harder than things that were farther away. The result would be the Earth and everything on it, is pulled into thin strings of atoms, a process technically called *spaghettification*, which sounds like a ride at Disneyland but is most assuredly not as fun—much cheaper, though. Once inside the event horizon, if you manage to somehow make it, you'd be in for a trip. Inside the event horizon, every direction points *down*, toward the center. Not even light can avoid this fate of being crushed to a tiny point. You would see literally nothing while you quickly descend to your ultimate fate of having all your atoms crushed to a point of infinite density.

Yikes. Though if that sounds like a shitty day, keep in mind that a black hole doesn't really need to collide with Earth to ruin your day. If it only gets close to our solar system, it can still fuck up the accuracy of the posters of space facts we put in classrooms by pulling planets from their orbits. If we are lucky, it'll only take Pluto with it so we don't have to read anymore Neil deGrasse Tyson tweets about its status as a planet. Things might be looking up after all.

The entire universe is going to decay into nothingness

Take a hot cup of coffee and a cold beer and set them on a table. Try not to drink them. We know. This could be the hardest thing you've ever done...

After thirty minutes or so, what happened to the drinks? Damn it—you drank them, didn't you? Okay, what would have happened if you left them there? Well, obviously, the coffee would get hotter and the beer would get colder. No. That doesn't sound right at all. That would never happen. But why?

The science behind this everyday fact is that of *thermodynamics*, which was invented in the mid-nineteenth century out of the practical concern of optimizing the efficiency of steam engines. If that sounds old-fashioned, well, yeah, we've come a long way since burning chunks of coal to heat water so that the steam moves turbines to create power. Huh...what's that? Oh, we still do this two hundred years later? For fuck's sake. Well... at least thermodynamics is still relevant.

The birth of modern thermodynamics is usually considered to be the publication of a slim volume called *Reflections on the Motive Power of Fire* in 1824 by Sadi Carnot, which sounds more like the album title of a British indie rock band. Though much of it was a precursor to what is now a very succinct set of "laws," it did contain one universal fact that will never be overthrown— useful energy can only be extracted when heat moves from something hot to something cold. This would later be enshrined as the *Second Law of Thermodynamics*.

In modern times, the Second Law is usually phrased in terms of entropy. *Entropy* is a famously confusing term. It is usually loosely associated with *disorder*. The more mixed up something is, the more disordered it is, and the more entropy it has. A messy room is a typical example. Leave a tidy room alone, and it will eventually become messy with seemingly no effort. However, to make it neat again always requires effort. The Second Law then states that *systems always tend to go from a state of low entropy to a state of high entropy*. This is slightly more general than Carnot's *hot-to-cold* version. By the way, Carnot was French, so you probably want to pronounce it correctly for your next dinner party.

Whether you think about entropy as heat or disorder or something else entirely, entropy is ultimately just about count-ing. We know you can handle it. Easy mode. How many ways can a room be clean? If you want it *perfectly* clean, then there is only one way. Ah, can you imagine it? A perfectly clean room even Marie Kondo could be proud of. What sparks joy? Low entropy, that's what.

Perfection aside, there are only a few ways a room can be

arranged that would convince someone that it is neat and tidy. That's easy to count. How many ways can a room be messy? Now *that* is impossible to count. But here's a better question: which state of the room would you call *the messiest*? This is where your counting prowess will shine.

Let's suppose there's a bookshelf in the room that holds ten books, and the state of the room is a single book on the floor. How many ways could that happen? Well, it could be any book. (It better not be this book!) That means there are ten ways. There are also many places on the floor it could be. There are ninety different ways for two books to be on the floor. You see where this is going. The messiest situation is when all books are off the shelf—there are just so many ways they could be arranged. In other words, the messiest situation is the one with the most number of ways it can arise. This is the *maximum* entropy situation.

A messy room is messy regardless of the details. Entropy counts the number of ways the details can give rise to the same general situation when zoomed out. The Second Law—that entropy tends to maximize—is just counting and statistics. If there are vastly more ways for a room to be messy and only one way for it to be clean, then a random state of the room will be messy. The Universe is like a room. But instead of large objects we can see, it is full of particles. Right now, it's like all but one of the books are on the shelf—many particles are clumped up in neat piles. You, for example, are a neat pile of particles (this is a great pickup line, by the way). But we must eventually end up with the mess—the particles spread more or less evenly across the Universe. The particles just go on doing their own thing. At any snapshot in time, they all occupy some state of the Universe. But the messy states—the high-entropy ones—vastly outnumber the states in which people exist—the lower-entropy ones.

Okay, but why does that matter? Can't interesting stuff happen in a high-entropy universe? No. Not only does nothing interesting happen; nothing happens at all. Contemplate that while staring into the deep blackness of your morning coffee. In fact, the coffee is a good analogy to explain why the end of the Universe is really the end. Suppose some hipster jackass drops a bit of soy milk in your otherwise perfect black brew. At first, it's just a small blob of white liquid set against the black liquid. In the end, of course, the entire cup becomes a smooth brown you can call a latte and sell for $7.50. This state was inevitable. If you think about the milk molecules, each individual one could be anywhere. Multiply that by the sheer number of them and you end up with countless ways to make a latte. But don't tell your barista lattes are trivial to make because of the Second Law of Thermodynamics.

The drop of milk is the lower-entropy state, and the latte is the higher-entropy state. In the beginning, "interesting" things happen to the milk. Note that "interesting" here is taken to be relative to whatever else you might be doing on a Sunday morning instead of staring at a cup of coffee. Ah, who are we kidding? There's never a bad time for an existential crisis. In any case, the milk spreads out in complex patterns you might even find at some contemporary art gallery. But it ultimately must turn into the latte, and you can stare at the latte all you want. Nothing will change at that point. It's the end. You'll never get that deliciously bitter black liquid gold back.

Our universe started in a simple low-entropy state. We are probably nearer the beginning than the end. We are like those patterns of milk in a cup of coffee. But the march to maximum entropy is ceaseless. The end of the Universe will be a cold abyss of empty space with the occasional lonely particle zipping by. And that will be the end of time. From the point of view of even those lonely particles, nothing happens. Zoom out and you have the latte universe, a cold weak soup of forever nothingness.

How can you tell we aren't in the latte universe? Easy. *Stuff* is happening. For stuff to happen, heat has to flow from one place to another. So already in the nineteenth century, scientists knew the Universe was going to end this way. Lord Kelvin called it the *heat death of the universe*. He didn't realize how confusing that was going to sound two hundred years later when scientific literacy would be at an all-time low. Of course, *we* understand that *heat death* is not *death by heat* but the death *of* heat, since heat needs to flow for stuff to happen. Don't worry, though, clever modern scientific communicators have renamed the concept to *the Big Freeze* in analogy with *the Big Bang*, which started the Universe, and some other end-of-the-universe scenarios, *the Big Crunch* and *the Big Rip*, which are also as bad as they sound (see Reason 36).

Is all of this depressing? Maybe. But there is something you can do about it. Well, only if you are one of those people that think "doing nothing" is something, because every action you take contributes to the heat death of the universe. If we go all the way back to Carnot, the real reason he wanted to study heat engines was to find out how efficient he could make steam-powered war machines. To be fair, this is back when war was just the constant state of the world. He found that heat flows from a hot place to a cold place, and the efficiency of the engine was related to the difference in those temperatures. In other words, every process must generate heat, equalize temperatures, and maximize entropy. If we are already at the maximum-entropy state, then of course we can't add more and nothing can happen. On the other hand, everything you do adds entropy to the Universe. The more effort you put into avoiding it, the faster you produce entropy and the closer you bring us to the end. So thanks for that.

The expanding universe could rip us apart atom by atom

Do you ever feel stretched too thin? Perhaps it's metaphorical, and you are just really bad at time management. Or the Universe is literally stretching your body apart from the inside like it's some torture technique from the Spanish Inquisition. Or maybe it's both, and you are being fucked by space and time simultaneously. How could the Universe allow this?

If you look out in the night sky, you'll see stars (with apologies to our friends in the perpetually cloudy UK). Every one of the thousands that can be seen on a clear night are only a fraction of the one hundred billion or so stars in our galaxy, the Milky Way. If you get your hands on a telescope, not only can you spy on your neighbors (probably a bad idea), you can see even more of the Milky Way's stars (definitely a good idea). Those new stars are farther away, but still in our galaxy. With even a shitty telescope you'd consider buying for your annoying nephew's next birthday, other objects beyond our galaxy can be seen. These are in fact more galaxies filled with billions of their own stars. We understand stars really well, and therein lies the problem—these faraway stars don't look quite right.

If you access a *research-grade* telescope, you'll see what several astronomers saw in the early twentieth century—that stars look too red. For them, it was like arriving home to find their partner sitting alone with a blank stare, but their face flushed bright pink. If the astronomer was Taylor Swift, we'd have yet another number one hit breakup song. Otherwise, yes, those red stars have some explaining to do. The quirk is that stars appear exactly the same in every direction, except that those farther away have all their colors shifted toward lower frequencies. For visible light, the lowest frequencies appear red, so we say that any light that has lower frequency than expected has been *red shifted*. When the opposite occurs, i.e., when light has *higher* frequency than expected, we say it is *blue shifted*, which is weird since indigo and violet are both higher in frequency than blue. But violet is actually hard for human eyes to see, and only graphic designers and color snobs would consider indigo anything other than an ugly shade of blue.

To understand the explanation for red shifted galaxies, it's easier to think about sound first. Why? Well, both sound and light travel as waves. Lower frequencies of sound produce lower tones, like bass blaring out of a car driven by an adult who forgot to grow up. Higher frequencies are those of screaming children and sirens, which together are probably not a good sign. The sirens are relevant though because they are the go-to example for something called the *Doppler shift*.

You may not have heard *about* the Doppler shift, but you

definitely have heard *a* Doppler shift. Chase down an ambulance, or try to remember the last time you saw one. Do you remember the sound of it zooming by? There's not really a word we can type for it, so we're kind of counting on your imagination here. When an ambulance is approaching you, the pitch of its siren seems higher. As it passes you, the pitch suddenly drops. The paramedics inside aren't changing the tone of the siren just for you—the sudden change is something only you can hear because of where you are standing.

When the siren emits sound, it sends out waves of specific frequencies. But the siren is also moving, so the waves in front of it get bunched up and the waves behind it get stretched out. This changes the frequency and hence the tone. Bunched-up waves are higher in frequency, and stretched-out waves are lower in frequency. You could say the sound of the siren moving toward you is *blue shifted* and the sound of the siren moving away from you is *red shifted*.

Now, with the powers of deduction, why do you think the light from all those galaxies is red shifted? Bingo! They are all moving away from us. What is more alarming, though, is that galaxies farther away are more red shifted. Everything is getting farther away from everything else. In other words, the space between galaxies is expanding. Nowadays, this phenomenon is comfortably described in modern physics as the expansion of space itself. Everywhere, at every moment, space is being created, pushing the existing space away. Think of it like the surface of a balloon that is being blown up. The rubber stretches out, and every point on the surface is pushed farther and farther away from every other point.

The weird thing about expansion is that it doesn't seem to have been constant throughout the history of the Universe. There was rapid expansion at the beginning—the Big Bang—and then little to no expansion until recently, where it seems like expansion is accelerating again. We can model it all very well with mathematics, but don't actually know why it happens. If we calculate the energy needed to drive this expansion, it seems to account for 70 percent of the energy in the Universe. We have no other way to detect it, so we call it *dark energy*. Ooo spooky! Ha ha. But actually, you probably should be a little bit scared.

The density of dark energy in the Universe seems to be constant. That means the amount has to grow since the volume of the Universe is increasing. But if the amount of dark energy changes, why can't the density of it *increase* as the Universe expands? We don't actually know, but it's not ruled out by our current laws of physics. So it's possible. In that scenario we call the energy *phantom energy*. It may seem at this point we are just ripping off a George Lucas script, but this is actual physics stuff.

The Universe is currently expanding. It is happening even inside of your body. Don't Google it, though, or you'll just be convinced you have diphtheria. In reality, the expansion is so slow that the forces keeping your atoms together don't even notice. In fact, the solar system doesn't notice. Not even Andromeda, the nearest large galaxy to our own, notices significant expansion and will actually crash into the Milky Way in about five billion years. The collision between two massive galaxies sounds awesome, but it would really be like throwing a couple of clouds at each other, which still seems like something cool might happen, but all you'll end up with is a bigger cloud. However, if phantom energy takes over, none of that will matter.

As the expansion of space gets faster and faster, the first thing that will happen is the disappearance of other galaxies. The space between them will eventually be expanding faster than the speed of light. Next, the stars in galaxies will start to spread as our own Milky Way dissolves. The spreading starts to happen more quickly as gravity can no longer hold the planets in orbit. The night sky will be black except for the Moon and our beloved planet. On the bright side, at least astrology will be dead. But soon everything else will be as well. In the darkness, the expansion will overcome the electrostatic forces holding the molecules in your body together as you—and the rest of the planet—are ripped apart. In the last fractions of a second, matter itself will be torn apart into quarks and other fundamental particles. After that, even space itself will be ripped apart. You can probably guess the name for this potential ending of the Universe. Yep, it's *the Big Rip*, which, if you accidentally read only that, would sound more like a surfing festival.

We know what you are thinking: *Even if I believe one of these crazy physics stories, you're still going to tell me it's a billion billion billion years away from happening.* Well...that all depends on how much phantom energy there might be. Once the increased acceleration starts to be noticed, there's very little time in any cosmological sense. To be sure, something else is definitely going to get us first (see Reasons 1, 2, 3...), but an estimate based on a particularly conservative model of phantom energy is only twenty-two billion years from now. We'd recommend getting a good seat, but it doesn't really matter where you are for this one.

PART VII

SO LET'S GO CRAZY

Sit down and get a drink.

REASON

You might turn into a ball of strange matter

You've heard this one before. Everything in the world is made of atoms. Atoms themselves are made of only a few things: protons, neutrons, and electrons. That means there is little difference between you and a rock. You and the rock are both made only of protons, neutrons, and electrons—they are just arranged in different patterns. These things make up *ordinary* matter, and it's really the only stuff we are certain exists (if anything exists at all—see Reason 39). This is probably not groundbreaking news for you since we teach children these facts, but it is not really the whole truth. There is other stuff out there that is not made of atoms, and by now you know if we're writing about it, this other stuff can't be good.

What you probably weren't told in school—because people stopped caring about teaching scientific facts to children long ago—was that protons and neutrons can be broken down further into three separate bits known as *quarks*. Quarks are the smallest bits of matter we currently know of. Now you're up to date. Everything is *not* made of atoms—everything is made of quarks.

There are only six types of quarks, which are referred to as *flavors*. Why flavors? No reason. Sometimes physicists get bored and give up when it comes time to name things. Speaking of which, the names of the quarks are *up, down, strange, charm, bottom,* and *top*. Again, there's no particularly good reason for any of the names. But we did list them in order of smallest to largest, which is important. Feel free to highlight them. You see, the Universe likes smaller things. Smaller mass means lower energy, and the Universe finds it easier to keep tabs on all its belongings when they are resting in a low-energy state. So we expect the Universe's favorite flavor of quark to be *up,* the quark with the lowest energy. It is, but just like ice cream, every once in a while, you need to try some stupid new hipster flavor like *tumeric and rose water with lemon poppy seeds* to remind yourself that chocolate is the only acceptable flavor, or up quarks if you're into that sort of thing.

So protons and neutrons are each made up of three quarks. The proton is made of two up quarks and one down quark, while the neutron is one up quark and two down quarks. Referring to the order above, this should mean that protons are lighter than neutrons, and that is indeed true. So protons have less energy and should be more favored than neutrons, right? Yes, that's also true. Protons outnumber neutrons in the Universe seven to one. There are lots of protons out there just happily flying around on their own, but a neutron on its own will turn into a proton (and some other stuff) within about fifteen minutes. This lowering of energy is called *decay*, and it's kind of unavoidable (which comes with its own depressing consequences—see Reason 35).

There are no *free* neutrons. All the neutrons in the Universe are bound together with protons to make larger elements—the things that appear on periodic tables. Remember that odd-shaped grid with random letters on the wall of every science classroom? That's the periodic table of the elements, and they are what make up everything you see around you. These elements exist in a state of *local stability*, which is just science-speak that means they are not easily broken up, but if they were, they could restabilize another way. Humans can test out this idea by applying a big enough jolt of energy, and thanks to science and engineering, we have increasingly bigger and more badass ways to apply this jolt.

The most extreme experiments we've created are called *particle accelerators* and, obviously, there's no point in accelerating something unless you plan on smashing it into something else. By accelerating particles, like protons for example, and smashing them together, we can not only find out what they are made of but also see all the other things that are possible—all the tools of cruelty and destruction the Universe can use against us. This is how we found quarks in the first place. First, we found the up and down quarks. Then, we found the strange quark in large particles called kaons and mesons and other "-ons." However, they are unstable and only last for tiny fractions of a second before decaying into regular stuff.

As awesome as these country-sized science experiments are, they have their limits. The relevant limit here is energy. The faster the particles can be accelerated, the more energy they will have, and the more stuff we can find. We may eventually produce enough energy to create large stable particles that are heavier than ordinary up-and-down quark matter. If that heavier matter contains strange quarks—the next heaviest after down quarks and therefore the most likely candidate—we call it *strange matter*. Perhaps the whimsical name of "strange" is appropriate because this stuff is truly weird. It's also really dangerous, but you probably don't remember it from your online occupational health and safety training—mostly because you weren't paying attention, but also because not standing in a bucket of water while working with live electrical wires is a far more pressing, though maybe equally unlikely, concern.

The simplest form of strange matter is made of an equal number of up, down, and strange quarks. This strange matter might be a state of local stability. That's fine. But if that were true, it would also be *more* stable than regular matter. That's a problem. Just as you might see in the process of crystallization (the growth of ice in your freezer, for example), one piece of strange matter can catalyze an unstoppable chain reaction of conversion of regular matter into strange matter. Down quarks would flip to strange quarks and be added to an ever-growing dense sludge of very stable strange matter until the entire planet became a single giant ball of strange matter. On the bright side, we've been called worse things.

But why would strange matter be more stable than ordinary matter in the first place? It's heavier, and so should decay back into a proton or neutron, right? Yes and no. Yes, eventually, everything will decay. However, *eventually* might be a really long time. Strange matter is potentially more stable than ordinary matter because it has more wiggle room. Protons and neutrons each have a pair of identically flavored quarks, so there are fewer possible ways they could come about. Think about making change when buying something with cash, as sketchy as that sounds circa 2023. If you have a $10 bill and two $5 bills, you have only $5, $10, $15, or $20 of exact change. But, if you have a $10, $5, and $1 bill, then you have $1, $5, $6, $11, $15, and $16 of exact change. Sure, it's less money overall, but it's a more stable change-making situation, which might come in handy if you have to make change quickly in a dark alley.

So strange matter takes more effort to reach but is very stable once attained. That means it would last a long time, and if it were created at some place in the Universe, then it's probably still flying around through space now. If it were to hit Earth or our Sun, that would be the end of life. So if we wanted to be specific, we'd stop telling children that they are made of atoms, and instead tell them that they are made of up and down quarks and if we added anything else they'd melt into a glob of strange matter. On second thought, atoms are fine.

The Universe could be deleted at any second

Everything decays—eventually, even you will be a frail, decrepit, lonely, withered old pale raisin, spoon-fed by your bitter progeny. But that's not what you should be worried about. You should be worried about the Universe itself decaying, erasing you faster than the search history on your dad's computer after he learned about private browsing.

The idea is called *vacuum decay*, and it has nothing to do with the crappy Hoover knockoff you brought from the shopping mall pop-up store because that slick salesman gave a demo showing how easily baking soda gets sucked off a doormat. Vacuum decay is about energy and how everything tends to end up in its lowest-energy state. Your smartphone doesn't spontaneously start charging itself. No, watching moronic TikTok videos only drains the energy available for your phone to do useful things, like stream YouTube videos of DIY furniture restoration.

The go-to analogy for anything energy-related is a ball

on a hill at the edge of a valley. At the top of the hill, it has a high amount of gravitational potential energy. And up there, it is unstable—any nudge will send the ball unavoidably rolling down the hill. The ball would make it up to the same height on the other side, then it would slip back down and up again—like a swing that never stops. This frictionless thought experiment is a favorite among physicists because they never had friends to push them on a swing. However, the real world contains friction and lonely physicists, one of which causes the ball to lose energy as it rolls. Eventually, it ends up at the bottom of the valley with no more energy to move. This is the fate of everything in the Universe—expending energy as useless heat in an endless march toward the lowest-energy state.

Lord Kelvin, the hugely influential nineteenth-century physicist, had a reassuring name for this: the heat death of the universe (and it's as awful as it sounds—see Reason 35). In this chapter, we'll make the heat death of the universe look like a boring weekend at your grandparent's Wi-Fi-less cottage. Okay, so here's the punchline: before the Universe reaches its final lowest-energy stable state, it might first spend time in a *meta*stable state. The Universe might even be in a metastable state right now! But calm down, because we haven't even told you what that is yet. A metastable state is a state that appears stable but isn't; it is a point at which everything seems fine, but might disintegrate at any moment.

The ball on the hill now sits happily forever in its valley. However, suppose that the valley is high up in the mountains. If the ball obtained some catalytic energy, like a gust of wind, it could make it over the next ridge and tumble down the entire mountain, releasing all that potential energy along the way. It might even hit other balls, kicking them out of their valleys. Eventually, there would be a mass of balls barreling down the mountain destroying everything in its path—kind of like an avalanche. Actually, it's exactly like an avalanche.

A metastable state posing as the vacuum state is referred to as a *false vacuum*. When something loses energy, moving from a higher-energy state to a lower-energy state, the process is called decay. So a false vacuum state decaying to the *true vacuum* is a process called vacuum decay. Hypothetically, facing down an avalanche of balls is kind of scary but not really worthy of Hollywood. Vacuum decay is the stuff of existential nightmares and definitely deserves at least a special on Netflix—obviously in Korean with English subtitles. [Reader laughing noises.]

Now before we calmly describe the complete annihilation of the Universe as we know it, perhaps you are already skeptical. Metastable. Really? That sounds like some theoretical mumbo jumbo. Is this just another wild idea from some cranky physicist who probably should not have gotten tenure? Well, actually, it's more familiar than you might expect. To see the total destruction of the Universe in a scale model, all you need is a bottle of pure water.

The pure water you will need is surprisingly easy to come by, given it generally dissolves everything. Of course, 100 percent pure liquid H_2O is impossible to find, but you can get close even with store-brand bottles. Oh, and don't believe what your Facebook acquaintance Rayne said about her artisan glacial melt mineral water—it's not "pure" or "natural," nor is whatever Gwyneth Paltrow suggests bathing in for your next water birth. Water will kill you anyway—see Reason 23 for more on that.

Once you have the bottle of pure water, stick it in the freezer.

Your freezer should be set to something like –18°C unless you are storing bodies, in which case we've found a slightly colder temperature is optimal. In any case, all the frozen shit clearly demonstrates that it is well below the freezing point of water in there. But even after several hours, you'll find your pure water isn't frozen?! That's right. Pure water does not freeze at –18°C— not without a catalyst anyway.

This is definitely something you should try, by the way. But we know you aren't going to get off your ass to do it, so we'll explain what happens. The pure liquid water will be– 18°C just like everything else in the freezer. However, it is not stable—it is in a *metastable* state. If the bottle is struck—by hitting it or hitting the bottle itself against something—it will start to freeze immediately! The solid H_2O is a lower-energy state, and this is where the water wants to be.

The water will start to freeze at a single point called the *nucleation* site and rapidly spread. Now, the whole thing would freeze solid except for the fact that as the water lowers its energy to the solid state, it must release the energy it had as heat. This energy increases the temperature of the water. So two things are happening to the water at once: it's moving to the lower-energy solid state and releasing heat energy as it goes. In the end, the bottle is filled with frozen water, but it is more of a slush than a solid chunk of ice.

This is like vacuum decay, but imagine instead that the water is the very fabric of space and time—which Einstein in all his brilliance underwhelmingly referred to as *spacetime*—and the resultant slush is the total destruction of the Universe. The analogy is apt in many ways. Let's start at the end. The end state is a completely different state of matter—solids have very different physical properties than liquids after all. It may be the case that after the Universe reaches its true vacuum state, the properties of physics in that state are completely different. And if they are different in any way at all, the chemistry will also be different. And if the chemistry is different, then all the matter in the Universe including the stars, planets, and water might cease to exist.

We don't really want to find out firsthand what some hypothetical true vacuum would be like. But it's not because the laws of physics might force our bodies to atomize. No. We don't want to visit the true vacuum because we would be disintegrated by the wall of fire traveling at the speed of light that brings it. Do you remember when the pure water rapidly decayed into a solid state? It started at some nucleation site and spread in all directions. Each water molecule dropped down to lower energy, crystallizing with the previous water molecule and releasing energy as heat to its surroundings. So what would the pure water that hasn't yet frozen see? Not a chunk of ice growing toward it—no, it would see first a wall of heat spreading like wildfire. The same would be true of vacuum decay—except you wouldn't see it coming.

Once some part of the Universe begins its decay, it catalyzes the decay in nearby locations, and this, in turn, catalyzes more nearby locations to decay. The process continues forever, spreading until the entire universe has reached the true vacuum state. But the decay is accompanied by a loss in energy. That must go somewhere—and that brings us back to the avalanche. Instead of a wall of icy cold air, the edge of the growing vacuum decay is a firewall of intense energy that essentially *deletes* everything in its path. But don't worry, it will be traveling at the speed of light, and so it's impossible to see it com...

We are stuck in the shittiest version of *The Matrix*

Have you ever played *The Sims*? Yeah, us neither. It looks super lame, and we only play cool vintage games on our classic Nintendos. The mechanics of *The Sims*, though, is to create a world that includes virtual people, called "Sims," and just watch them do stuff. There's no "point" to the game and no way to "beat" it. The Sims just go about their little virtual lives doing what kind of looks like normal human stuff. Of course, if you are a pimpled and horny teenage boy, you probably force your Sims to do...well...other stuff. Now, that's all in good fun...except for one small thing—you probably are a Sim yourself! Plot twist!

Even if you haven't played *The Sims*, you are probably still familiar with *The Matrix*, a 1999 science fiction film starring a group of people dressed like they weren't sure whether they were attending a wedding or a BDSM sex party, so they tried to dress for both at the same time. In the film, the protagonist, Neo, discovers that Earth has been taken over by intelligent machines (very plausible—see Reason 10) that have created a virtual reality world, the Matrix, to distract humans while using their bodies as batteries. It's all very late '90s macabre—it was *The Sims*, but for the *angsty* Gen X crowd...who just wanted to throw sex party weddings. Also, it's not just science fiction—you are probably in the Matrix yourself! Double twist!

The Sims and *The Matrix* seem completely different on the surface, but they have one crucial thing in common. There is a "real" world of intelligent beings that control a "virtual" world of intelligent beings who *think* their virtual world is real. Read it again. As it was for the humans "in" the Matrix, a realistic simulation would simply require all sights, smells, and feelings one experienced to be "uploaded" into a sleeping brain. This would give the perception that the person was moving around in a "real" world that exists outside of their inner thoughts. Spare a thought for the being that had to digitize the taste of vomit or the smell of burned hair for our unconscious minds.

From the perspective of those in the "real" world—the one outside the Matrix—it is obvious that those in the virtual world are being fooled. But what if the "virtual" world included computer simulations of even more virtual worlds? Think a new Matrix *in* the Matrix. Those virtual computer programmers—the ones in the Matrix—would look at their metavirtual creations—the ones in the Matrix in a Matrix—and be even more convinced they inhabited the "real" world despite living in yet another fake virtual world. What a tangled fucking mess. Let's reiterate, and start from a highly hypothetical beginning.

You are pretty smart—we can tell—so smart, in fact, that you've created a computer program that can simulate a human brain. Go you. In your simulation, your virtual brain creation has all the features of a real brain, including the illusion that it perceives a world full of things it can touch and feel. One of those things might be a computer. So your creation might fancy itself a smart cookie and program another simulation—a simulation in a simulation, as it were. And so on it might go, virtual worlds inside virtual worlds, all filled with simulated intelligences that think they are at the top of the chain. They each think they are *you,* the first mover, the first programmer in the one true reality. It's as if Christopher Nolan directed the Universe.

It's cute that you think you are it, the top dog, the prime programmer. Take a step back though. If there are simulations of simulations of simulations of...then...well...there's going to be a lot of porn. But more to the point, there are also *way* more simulated beings than not. So, odds are, you are definitely not the *one* person at the top of the chain. It's just a numbers game. There are potentially shit tons of people (or conscious *things*). One of them is the original programmer living in the physical world. Countless others exist in virtual worlds. If you asked yourself, *Which world does some randomly selected perceiver inhabit?* surely the answer would be a virtual world. In other words, the logical conclusion about any thinking entity is that it exists in a simulation—it is, and you are, *in* the Matrix (with very high probability). This is referred to in philosophy circles—which we aren't invited to anymore—as the *simulation argument.*

Okay, so we are all in a simulation. That's cool. But wait; if that were really true, it'd be a pretty shit simulation. Clearly this version of reality simulation software sucks. It's like a bootlegged copy of Windows 95 running on an iPhone 4. We have no idea what that means, but it sounds like a shit experience. Maybe that's the point though. See, to simulate a universe inside of a universe is difficult, at least if you want to do it in very fine detail. However, you don't really need to do that if you want to simulate only the *experience* of reality. You need only look around at this shithole to see—or, more appropriately, not see—that.

Look around you right now. What do you see? Tables, chairs, doors, your closet full of leather from your wedding... in other words, *stuff.* But you already know that is an illusion. That stuff is all made of atoms with mostly empty space between them. And those atoms are made of even smaller things and more empty space. We don't *see* any of that. Even our sense of free will is an illusion according to neuroscience gurus. The details, which would require far too much computational power, need not be simulated if all you want to program is a bunch of fools. We only see what we need to see to survive and procreate—or whatever the programmers of our simulation want us to see!

Let's pause for a moment...get a drink and take stock. Sure, a future technologically advanced civilization might have the computing power to run super realistic simulations of other intelligent beings if they wanted to. But maybe they don't want to. Maybe they voted for decent leaders—instead of shit career politicians or corrupt wannabe demagogues—who have passed laws against the unethical practice of simulating conscious entities. Or maybe we voted for the usual suspects enough times that we eventually ran ourselves into extinction. One of these possibilities seems implausible, and the other is just depressing. So the point remains: if civilization lasts long enough to be able to simulate conscious entities, then most of the people who will ever live to experience stuff will do so in a virtual world.

Don't think this is some future problem either. Just like grandpas who can't get enough World War II documentaries, one thing a future civilization might want to do is simulate human history. Perhaps, then, we are just here for entertainment or some historical science experiment run by future high school students. So let's put on a good show, fellow puppets! Otherwise, our pimply-faced overlords might pull the plug. We'll bring the leather.

REASON

You might have a mini black hole in your pants right now

Life is pretty complex. That's not meant to be deep or anything—it's just a cold hard fact. But we aren't surprised by the intricacies of life since we are so familiar with them. There are some strange things in the Universe that we know exist but are impossible for us to comprehend. In some sense, they are simpler than life itself, yet you wouldn't want to touch them. We've actually encountered them a few times already (there's more than one reason to hate them—see Reason 34). They are *black holes*.

Black holes are curious things. We only directly observed them in 2017, with the Event Horizon Telescope. However, we already suspected they existed as far back as 1915, when they fell out of Einstein's equations in his general theory of relativity. While Einstein will always be remembered as a zany-haired celebrity genius, there was a time when he was an obscure nobody. He spent nearly a decade developing his theory by learning an entirely new form of mathematics that had never been applied to physics before. Then he ended up in America, and his career went to shit. Ah yes, the American dream. But let's keep our rose-colored glasses on—this is a book about optimism!

In Newton's physics, which for hundreds of years were considered to be the true model of reality, stuff obeys laws of motion, which prescribe how it moves in the three dimensions we call *space*—up-down, left-right, and back-forth. *Space* is just the name for the three coordinates we need to locate something, not *outer* space, which is where we send astronauts and billionaires—not permanently, sadly. For Newton and all physicists up to the twentieth century, space was absolute—a fixed backdrop where the events of the Universe played out. Time also was absolute. There was some (at least hypothetical) clock that could be referenced by anyone, anywhere that displayed the universal time. Indeed, Newton's vision of the world is sometimes called the *clockwork universe*. Einstein took the clock and fucking smashed it.

For Einstein, space and time are not separate things. He showed us that there is a single four-dimensional object called *spacetime*. For Einstein, spacetime is also not absolute. It changes, bending and curving in response to energy and matter. Most importantly, as Einstein conceived it, spacetime is not the same for every observer. Space and time are *relative*. That last one is a mindfuck. Luckily, it's not that important to understand why black holes exist and why they can really ruin your day. So we aren't going to worry about it. Why bring it up, you are wondering. To look smart, obviously.

Based on the aforementioned principles, Einstein wrote

down some wicked equations that relate the *bending* of space-time to the amount of mass and energy nearby. Not a month later, Karl Schwarzschild used Einstein's equations to figure out precisely how gravity worked around a completely stationary, perfect sphere. These objects are very rare in the Universe, but this was a big deal because it opened the door for making other predictions. Things were really looking up—not for Schwarzschild though. He died less than a year later. There was also a problem with his solution. If the size of the object was below a certain limit, then a *singularity* appeared in the equations. A singularity is just fancy mathematics jargon for dividing by zero, which you'll remember from grade school is a big no-no. Depending on how old you are, your parents would have been hit with a stick if their mathematics teacher found them dividing by zero. Ah, the good old days...

Now, usually, when a singularity appears, it signals something has gone wrong. Dividing by zero produces infinity. In the physical world, there are no infinities, except doomscrolling endlessly through Twitter. In Schwarzschild's solution, he suggested that a mass that was dense enough would produce infinite curvature of spacetime. To most at the time, this simply suggested a failure of the solution in that scenario. So not many physicists, including Einstein himself, took it seriously. On the bright side, the mass density required for this to happen was ridiculously high. The Sun, for example, weighs in at two nonillion kilos. Yeah, *nonillion* is a word. In plainer but no easier to understand terms, it is two thousand billion billion billion kilograms of mostly hydrogen and helium. It's also really old too, if you want to be rude about it. On the other hand, the radius of the Sun is 700,000 kilometers. This actually works out to about 1.4 kg per liter, or about one and a half pineapples shoved into a space about as big as a wine bottle.

In order for things to go wrong in Schwarzschild's solution, the Sun's radius would have to be only 3 kilometers, giving it the density of your average TikTok influencer's skull, or a quadrillion times the density of the densest substance on Earth, which, by the way, is osmium. The density of an enormous black hole is actually less than that of water but requires an amount of mass a billion times that of the Sun. In either case, scientists just didn't believe such extremes were possible. Turns out they were wrong. Black holes are plentiful in the universe and come in a variety of masses. If nothing else, the Universe is a big fan of diversity.

For a given mass, the radius that Schwarzschild found is now called the *event horizon*—hey, just like that telescope… what a coincidence! As more mass enters the sphere defined by that radius, the black hole grows in size. For a long time, it was assumed that crossing the event horizon was a one-way ticket. You may have heard that nothing—not even light—can escape a black hole (hence the name). But in the 1970s, Stephen Hawking proved that black holes have a temperature and thus radiate heat, now called *Hawking radiation*.

The interesting thing about Hawking radiation is that it too was assumed to be very weak. For example, the temperature of a black hole the mass of the Sun would be in the nano degrees or, more technically, cold as fuck. The larger the black hole, the colder it will be. And black holes are generally assumed to be large. There are basically three types of black holes. *Supermassive black holes* are created when smaller black holes suck up stars or merge with other black holes. These are truly monsters. *Stellar mass black holes* are created when stars go supernova (not as fun as it sounds—see Reason 31). These more common black holes can be real shitheads (see Reason 34). And, last but not least, the hero of this chapter is ready to make its debut—the deceptively named *mini black holes*. These are the black holes with small masses, like the weight of a marble. We know super-massive black holes and stellar mass black holes exist. No one has seen a mini black hole. But here's the thing: you probably wouldn't want to see one either.

How could you find a mini black hole in the first place? Well, first of all, they are black, so it's not like you are going to spot them with a telescope. That's also true of regular black holes... unless you have a telescope the size of the Earth and an army of poorly paid graduate students. We usually find black holes by observing their effects on the stuff around them. They typically have a lot of mass—much more than our Sun—and so will have things orbiting around them. If you find a bunch of stars orbiting around something very heavy but find nothing there...good chance it's a black hole. But mini black holes are small, so they are not going to affect the space around them much—unless of course there is stuff really close by, which brings us to the point. The only way you are going to find a mini black hole is if you end up with one in your pocket. But if that happened, you wouldn't even get a Nobel Prize for it—because those are only given to living people...and most people don't survive their skin being ripped off and bones crushed. Well, that got your attention.

Suppose you reached into your pocket and pulled out a black hole. If it were the *size* of a marble, it would be the weight of the entire Earth. Good thing you didn't skip leg day. But two Earths pulling on you will only make you *feel* twice as heavy, right? Wrong! You must remember that you are very far from most of the mass of the Earth. The *surface gravity* is the force experienced on or near the surface of an object. On Earth, it is defined as 1 g and is a great reference for other forces you might experience. For example, on the Moon, you would feel 0.16 g or 16 percent of Earth's surface gravity. In free fall, you feel 0 g. On the "surface" of the Sun, you'd feel 28 g, which would definitely not be pleasant. You may be wondering, *Just how many g's could I handle?*

A typical human, which would statistically be a thirty-one-year-old female named Mohammed Wang, can withstand about 5 g for no more than ten seconds before losing consciousness. Now, how about that black hole? The surface gravity of our marble-sized black hole is 5 quintillion g. You'd basically be immediately ripped to shreds. Then the black hole would start to consume the Earth from the inside out. Anything not swallowed by the marble would end up a dust ring as the tidal forces pulled everything apart. Nothing would survive, except, of course, Keith Richards or those little water bear things. In any case, please don't pull a black hole marble from your pocket.

How about a black hole with the *mass* of a marble? First of all, it would be tiny—like less than a trillionth of the size of an electron. So taking it out of your pocket is going to be difficult. But no matter. You'd be dead as soon as it appeared. Remember Hawking radiation? Well, if something radiates, it releases energy. If it releases energy, it loses mass. So according to Hawking's calculations, black holes must lose mass if they aren't sucking any in. But almost all of the energy is released in the final moments. In other words, tiny black holes are miniature atomic bombs. Your marble-mass black hole would last for a length of time so short we don't even have a name for it. Let's just say it's instantaneous. And yeah, it would explode in a flash with the power of a few nuclear bombs. You, and everyone around you, would be vaporized (more on what that would be like in Reason 8—spoiler: it's not good).

So, like, is a mini black hole going to show up in your pocket or what? Probably not, because the energy required for one to be created would be the same as that released when it goes pop. And that energy would have to be concentrated into an impossibly small size. There is no conceivable technology that can accomplish this, nor is it going to happen spontaneously... at least not today. There could be mini black holes from the distant past scattered throughout the Universe. If you've heard anything about Big Bang theory (either the actual physics thing or the inexplicably more popular sitcom), then you'll know the Universe started very hot and very dense. Some models suggest that mini black holes, called *primordial* black holes, could have been created during this brief time at the beginning of the Universe. If they were the mass of a mountain to start with, they'd have survived long enough to still be here nearly fourteen billion years later. There could even be lots of them. We could run into one, or one could run into you. There might even be one in your pocket!

REASON

Dark matter killed the dinosaurs…and you're next

It's been a long journey to the penultimate reason to hate this godforsaken place, so let's get straight to the point: about 85 percent of matter in the Universe is of unknown origin. Eighty. Five. Percent. Imagine you are a teacher on an excursion and you've returned with only four and a half out of the thirty kids you left with. Disappointing? Perhaps. But 15 percent is better than nothing. On the other hand, you know the other 85 percent, possibly fewer, are going to turn up at some point with less than savory tidings.

This 85 percent of unaccounted-for matter is uninspiringly called *dark matter—dark* because we can't see it. We know it's there because it is heavy and pulls with its gravity on entire galaxies, keeping them together, and we like galaxies. They are big, bright, and beautiful structures, and photoshopped images of them inspire every generation of astronomers, astronauts, engineers, and megalomaniac CEOs. But get this: galaxies aren't bright *enough*. The number of stars just isn't enough to explain how they stay clumped together. You may be rightly wondering, *How the hell could we possibly know this?*

We see lots of stars spinning around the center of gravity in galaxies. From these observations, we can get a good handle on how much mass a galaxy has. This tells us how strong the force of gravity is. The problem is that the mass of the visible matter alone doesn't produce enough gravity to hold the galaxy together. All those stars should just fly off into the void. Now if the calculation were off by just a little bit, then maybe we'd accept that a measurement or two might be faulty. But we are missing 85 percent of the matter that seems to be producing all that gravity. It's kind of embarrassing, really.

Now, you're clever, so you must be thinking there are plenty of things of mass that don't emit light. How about clouds of cold gasses, black holes, or even antimatter clusters? No, sorry, these would be obviously detected. Scientists have ruled out every alternative that exists within our current understanding of physics. Fine. But if this dark matter is the vast majority of matter, then it should be everywhere, right? That is, here on Earth too, and even inside of our bodies. And, since no one has shown up at any hospital for dark matter injuries yet, it must be pretty safe...right? Ummm, yeah, that's a solid maybe.

Even though we've got no idea what dark matter actually is, you could think of it as a huge cloud of dust and gas that clumps around galaxies. Now, unless you are a cosmologist, we can pretty confidently say that the picture you currently have in your mind is wrong. You are probably imagining a cloud of gas as if it were as thick as a puff of cigarette smoke from a leather-faced

Boomer, back when it made them look as cool as Keith Richards on a perpetual bender. This is all wrong. The actual density of our galaxy is tiny, and so the cloud of dark matter is wispy, like that left behind an hour after a teenager had their first drag from a vape on a windy day. To put a finer point on it, if you took a Super Big Gulp–sized cup and scooped up a sample of the Milky Way, your cup would contain a few thousand atoms. Yet, when full of Fruity Watermelon Lemon Fiesta, your supersized cup contains a billion billion billion atoms.

To get a real sense of just how sparse the galaxy is, consider that the best vacuum chambers in the most technologically advanced laboratories in the world are still ten times the average density of the Milky Way. So even if there is five times as much dark matter as regular matter, its overall density is pretty low. Perhaps then, we don't need to worry about it. Although...the density of regular matter is pretty high in *some* places within the galaxy. That is to say, even if an average patch of the galaxy is boring, we live in a pretty remarkable place. Well, those of us outside of southern Idaho anyway. What if dark matter could clump together in a similar fashion? There are a couple of ways scientists have proposed this could happen. As you might have expected, neither bodes well for us. In an actual peer-reviewed journal article titled, "Death and Serious Injury from Dark Matter," the authors boldly conclude, "Our results open a new window on dark matter: the human body as a dark matter detector." Any volunteers?

One theory of dark matter suggests that it consists of very dense and largish objects that may interact with regular matter like the flesh in the sacks of meat we steer around using our brains. Exactly *how* it interacts is not important to answer the fun questions. All we need to assume is that energy is conserved, which is something we tell children in YouTube videos, so it must be true. If you didn't like and subscribe, the punch line is that the collision will cause the dark matter to lose energy, and that amount must go somewhere—namely, into us.

Now, if you calculate how much energy is deposited into a body that has dark matter passing through it, you'll get a number that depends on the size and speed of the dark matter. Right away, lots of combinations can be ruled out based on years of experiments and observations. However, for dark matter that is about as heavy as a large child packed into the size of a bacterium, the human body is a great detector. Bullet wounds would have been the obvious analogy, but we are artists.

If dark matter is just sitting around in the galaxy, it's not as if it is going to be ramming into us. But wait. Our solar system is actually traveling about 250 kilometers every second around the center of the galaxy. That's over seven hundred times the speed of sound and five times faster than anything ever made by humans. So if we think of ourselves as sitting still, then all the actually still stuff in the galaxy is indeed flying at us like bullets from the wildest dreams of sci-fi nerds. Let's reiterate. There may be chunks of dark matter, too heavy to put in your carry-on luggage, barreling toward you at cosmic speeds. If that sounds terrifying, don't worry; it gets worse. The amount of energy left behind as the dark matter bullet streaks through you is enough to heat your flesh to over 10 million degrees Celsius. This would immediately create a vaporized hole wide enough to see straight through. Now, no studies have been done to evaluate the effects of such extremes on the human body, but we are presuming death is a likely consequence.

As expected, the odds of you being sci-fi laser-gunned to death by dark matter bullets is small. But then again, so is winning the lottery, and most people seem to expect that to happen at least once in their lifetimes. However, even if dark matter doesn't kill anyone *directly*, it might be the cause of something much, *much* worse. Mass extinctions...ever hear of those? (See Reasons 5 and 6 for why you are probably enabling one right now, by the way.)

Mass extinctions are surprisingly common—it's not just the dinosaurs that spontaneously disappeared from the fossil record. The most surprising thing, though, is there seems to be a cosmic scale pattern to them. Geological evidence suggests a massive loss of species diversity every thirty million years. Here's an interesting sidebar: our solar system bobs up and down as it goes around the galaxy. How often does it cross the plane of our galaxy? Hmmm...once every thirty million years. Oh, and how often is there a peak in crater formation in the geological record? No, it can't be. Yes, it can. It seems that impact events also happen on a periodic thirty-million-year timescale. Of course, this could all be a coincidence, but perhaps it is not.

Our galaxy is large and disk-shaped, with stars clustered mostly in two main spiral arms. It is remarkably thin—a cosmic-sized CD, if anyone remembers those. Our solar system, the Sun and Earth and other stuff, spins around inside the CD like it's playing out some sweet-ass '90s music like Soundgarden or Celine Dion. Like the bumps and grooves in a CD, the solar system also bobs up and down as it goes, crossing the midplane every thirty million years. If dark matter also clumps in this disk, then it might shake up the trillions of comets floating around in a cloud around our solar system as we pass through it. It would be like running into a thin wall of compressed air. You can't see it. You'll make it through mostly unscathed, but it's going to mess up your hair, and that might send a comet off course hurtling toward you. As cosmologist Lisa Randall wrote in bestselling fashion, dark matter might have killed the dinosaurs. Wait. Was that about sixty million years ago? Shit.

We are an insignificant speck next to the scale of the Universe

The observable universe is one hundred billion light-years across. This is the largest thing we know exists. The smallest things we know exist are quarks, which are about ten billionths of a nanometer. These are just words, so let's try to understand just how big this difference in scale is.

The first time you encountered a logarithmic scale was probably in an elementary school science class, thanks to the wonders of litmus paper. The quantification of acidity came in the early twentieth century from beer scientists. No, that's not a typo. At the Carlsberg Laboratory, Soren Sorenson Jr., a Danish chemist, discovered that hydrogen ions were the key to acidity...and presumably good beer; this led to the pH scale. So *pH* stands for "power of hydrogen" and is a numerical scale from 0 to 14, with smaller numbers being more acidic. Smack-dab in the middle is the number 7, which is the pH of pure water. The most important thing about this scale is that it is *logarithmic*, which means that every number is really a tenfold change. So your pee, with a pH of 6, is ten times as acidic as water, but that's not why you shouldn't drink it. Stomach acid has a pH of around 3, making it ten thousand times as acidic as water. Something like battery acid, with a pH of 1, is a million times more acidic than water. That's the power of logarithmic scales. A softer way to phrase the difference in the levels of a logarithmic scale is with the terminology *orders of magnitude*. So, something that is a million times more acidic than water is said to be *six* orders of magnitude more acidic.

All right, cool—now that we are all calibrated, let's return to the scale of the Universe. An average human can stretch their arms out and reach 1 meter. That's the limit of your immediate influence. You may be reading this in a cozy one-bedroom apartment that is 10 meters across. That would be about one order of magnitude larger than you can reach—so spacious! A football field is 100 meters long and two orders of magnitude longer than your feelers. The Gigafactory Texas, built by Tesla to build Teslas, is just over a kilometer long. It is three orders of magnitude longer than you. The longest thing humans have ever built is the Great Wall of China. At over 20,000 kilometers long, it is seven orders of magnitude longer than your reach. It's only a fifth of the circumference of the entire Earth, so that's about it for the size limit of what we could construct or ruin with our clammy paws. We have traveled much farther to ruin shit though.

The farthest anyone has traveled from home was to the Moon, where we did donuts in cars, planted flags, hit golf balls all around, and left a bunch of trash and bags of our own shit (true—see Reason 29). So in terms of our destructive influence, we seem to be limited to about eight orders of magnitude larger than ourselves. Not bad. But there is so much more out there we could contaminate that we're missing out on. For example, the Sun is eleven orders of magnitude out of grasp for those clammy, five-digit tools of masturbation on the end of your arms. We've smashed some shit into the Sun, but it didn't leave a dent. Vengeance is probably still on the way though (see Reasons 28 and 33).

We have limited influence beyond our solar system. However, we have sent probes flying off into outer space. The Voyager 1 probe is thirteen orders of magnitude farther away from us than our outstretched arms can reach. Perhaps it will run into a small rock, and we can celebrate a new limit to our antics. But that's definitely it. Beyond our solar system is just empty space for many lifetimes' worth of distance we could cover. The nearest star is over sixteen orders of magnitude beyond our reach. The nearest galaxy is twenty-two orders of magnitude beyond our tickling limit. Finally, the observable universe is twenty-seven orders of magnitude bigger than our puny little meat sacks. You just have to pause and appreciate that fact. There is no analogy relevant here. We are insignificant next to the vastness of the cosmos. But before we get all poetic about it, let's also consider the other direction.

At the end of your scrawny appendages are your grabbers. These marvels of evolution and dexterity that most of us use to pick body secretions out of various cavities are one order of magnitude smaller than the whole arm. Penis size jokes aside, the smallest thing you might be able to manipulate with your hands is only two orders of magnitude smaller than you—maybe the tip of a tiny pencil. Speaking of which, the world record for smallest handwriting is another order of magnitude smaller than that—a few letters written on a grain of rice. It is readable with the naked eye, but we can only see things four orders of magnitude smaller than our bodies, so your visual acuity is definitely not the reason you or your partner (or whoever else you've invited) can't find their penis. Beyond that, we need tools to see. With a light microscope, we can see our individual cells and the tiny creatures they are in constant battle with (see Reason 15 for the sitrep). The limit our aided eyes can see is seven orders of magnitude smaller than we are. We can reach out and grab some moondust and then see the flecks of it under our fingernails. These are the limits of even our aided senses. This range spans fifteen orders of magnitude.

Computers allow us to go even deeper. The smallest transistors in computers now are about nine orders of magnitude smaller than our body. Stick a bunch of these things together and attach them to an electron microscope, and we can see individual atoms, which are almost ten orders of magnitude our junior. The proton at the center of that hydrogen atom is fifteen orders of magnitude smaller, and this is the limit of what we know for sure. Well, sort of... We're pretty sure quarks exist (which could be a bad thing—see Reason 37), which are estimated to be about twenty orders of magnitude down. Beyond that is speculation, which you can definitely find a lot of in theoretical physics, but we've gone far enough to make the point.

There is no doubt we are impressed with ourselves. We have made a lot of progress by increasing our influence a few orders of magnitude in either direction with inventions such as politics and war aided by science. But this limit is really it. Considering the relatively paltry limit we have reached largely involves us smashing shit into other shit, perhaps it is wise of the Universe to have contained us.

So here we are, sacks of elementary school pH experiments propped up by brittle calcium sticks flailing our noodle-y protuberances in vain attempts to spread our filth beyond our reach. From quarks to the whole shebang, the known universe spans forty-six orders of magnitude, and it's probably much bigger. With our clumsy meat hooks, we have direct access to only a tiny sliver of it. Technology has aided us in extending our reach, but at best it has only shown our depressing limitations in very certain terms.

The one reason **not** to hate the Universe

Congratulations. Since you're reading these words, it means you've either read the whole book or skipped ahead to find that one reason to love the Universe. Either way, we hope that in this chapter, you find inspiration and motivation, something that puts a little bounce in your step as you journey through life. Despite all the previous chapters making it seem foolish to consider any reason to *love* the Universe, especially considering how fleeting these times of positivity are within all the suffering, inequity, and short-sightedness in this cosmos. But out of the darkness, a simple thought presents itself. You'd better sit down.

As has been made clear throughout this book, the 42 reasons to hate the Universe give us a glimpse into the harshness of the cosmos. However, we implore you to take another perspective. What if the reason to love the Universe *is the* 42 reasons to hate it? We know. We're sorry, but hear us out. You are the product of billions of years of cosmic and biological evolution.

And—although you're now sitting with your pants unbuttoned and wiping the potato chip crumbs off your chest, aware of the impending struggle to get up off the couch as you near the end of this book—you *are* a miracle of existence. From a cosmic perspective, the couch you are struggling to get up from is sitting on the surface of a rock that is perpetually falling around a ball of continual nuclear explosions. We are just far enough away that its bright glow and gentle warmth have bathed the Earth for billions of years, allowing life to flourish—only to be shut out as you close the blinds because it's reflecting off the TV, blocking out who is currently being voted off *Love Island*.

But all that is only mind-boggling; what is downright unbelievable is the chain of events that brought you to this point. The Universe had to first burst into existence in a chance event no one knows how to explain. Fourteen billion years ago, when the Universe was the size of a pea, fluctuations had to create a subtle difference in energy density to form any structure at all. We had to wait hundreds of millions of years for gravity to do its thing, pulling hydrogen together and creating the first stars. Then, we waited some more for those stars—the ones that were massive enough, anyway—to explode in a very particular way, sending the carbon and other heavier elements forged in their hearts to the right places, only to be pulled together again. This cycle had to be repeated several times to create just the right proportions of elements so that our own solar system could spin into existence and form the perfect neighborhood for life to flourish.

Earth is one of several planets in our system, and we are lucky enough for it to be perfectly positioned inside the habitable zone of our star, the Sun. The Sun was literally blasted into existence from an older, much larger star that went supernova, creating the supermassive black hole at the heart of our galaxy as well as all the other stars in the Milky Way. Our galaxy is part of a smallish collection of other galaxies known as the local group, which itself is part of the Virgo Supercluster, including thousands of other galaxies. The Virgo Supercluster makes up less than 1 percent of the observable Universe. We are truly a speck of dust in an empty Universe.

Of course, our precious Earth didn't form in some single glorious moment of creation—it was a hellscape of fire and death in the beginning. In fact, another planet, *Theia*, crashed into the early Earth, sending a giant chunk of it off at just the right speed and angle to be trapped in orbit, creating our Moon, which in turn causes the tides, lights up the night, and facilitates the progress of life for many species. Even in the billions of years following the creation of the Earth and Moon, our planet has been struck by just the right number of other asteroids, sparing the lives of every one of your ancient ancestors while eliminating much of their competition. But those ancestors still had to endure countless catastrophes and near misses.

There is no hiding the fragility of life on Earth, and we humans are not exempt. Recall that a single geological event nearly one hundred thousand years ago was enough to reduce the human population to only a few thousand individuals. There have been no signs of life anywhere else in the Universe. We crave the comfort of being surrounded by our family, friends, and even sometimes strangers. Our loneliness has led us to reach out far beyond our solar system. But nevertheless, the spin-offs of these ventures into the cosmos have given us technologies that have brought us closer together and allowed us to live much longer. So much so that everyone alive today probably has some form of technology to thank for the preservation of their own life or the life of an ancestor.

In more modern times, leading up to your birth, your grandparents and even parents likely dodged the effects of war, plagues, diseases, famine, and many other terrestrial threats. And you will likely survive many more chance encounters—perhaps even procreating to give yet another human being a chance to overcome the unfathomable odds faced by any organism lucky enough to live on this planet. Think about how one single insignificant event in a distant ancestor's life could have led to our species residing in the darkness of obscurity for all eternity. What if the cataclysmic events that plagued Earth's previous inhabitants never came to be, leaving our ancestors no chance to take hold and cement their place as a dominant force in the rich history of our planet? Humans may never have had an opportunity to flourish, learn, and engage with the world around us. Think of all the minuscule probabilities that have fallen in favor of this very moment—from chance meetings to the avoidance of illness, injury, and death.

And this brings us back to *you*—you began in two places. In one place, a cell will have gone through several stages of development, competing with half a million other cells only to lie in wait for the other half necessary to form the unique DNA instructions to make you. Your other half may have formed years later and likely thousands of kilometers away. There, a tiny germ cell divided and changed until it resembled a tadpole. Then, together with 500 million other tadpoles, one-half of you was pushed—probably prematurely—through a semi-flaccid penis in a moment that was likely anticlimactic for at least one person, but a miracle for you.

You're an improbable combination of trillions of cells, each with the same unique DNA instruction manual. The same DNA that dictates the color of your eyes, the shape of your eyebrows, how you metabolize a glass of wine, your reaction to a bee sting, and whether you'll get Alzheimer's at sixty-five or seventy-five. It's easy to forget that these same instructions are also responsible for the cellular reactions that make you feel love, sorrow, happiness, and fear. (And if you've binge-read this book, you're probably feeling a bit terrorized right now.) You are an ever-changing combination of memories, experiences, and information you've gathered in a life that will only exist for the briefest moments on a cosmic scale. Yet your life is the result of billions of years of stardust that has evolved and changed to culminate at this moment. Sure, those processes created every doomsday scenario we've written about (and a great deal more), but they also created you. You are a part of this shitshow, and it's time to own it. Love this universe because you are it, and it is you.

BIBLIOGRAPHY

REASON 1

Boeree, Liv. "Why Haven't We Found Aliens Yet?" Vox, July 3, 2018, last modified April 13, 2022. https://www.vox.com/science-and-health/2018/7/3/17522810/aliens-fermi-paradox-drake-equation.

Bryson, Steve, Michelle Kunimoto, Ravi K. Kopparapu, Jeffrey L. Coughlin, William J. Borucki, David Koch, et al. "The Occurrence of Rocky Habitable Zone Planets around Solar-Like Stars from Kepler Data." *The Astronomical Journal* 161, no. 1 (2021): 36. https://doi.org/10.3847/1538-3881/abc418.

Hanson, Robin. 1998. "The Great Filter—Are We Almost Past It?" September 15, 1998, last modified April 17, 2017. https://mason.gmu.edu/~rhanson/greatfilter.html.

Martin, William F., Sriram Garg, and Verena Zimorski. "Endosymbiotic Theories for Eukaryote Origin." *Philosophical Transactions of the Royal Society B: Biological Sciences* 370, no. 1678 (2015): 20140330. https://doi.org/10.1098/rstb.2014.0330.

Sheikh, Sofia. "Nine Axes of Merit for Technosignature Searches." *International Journal of Astrobiology* 19, no. 3 (2020): 237–243. https://doi.org/10.1017/S1473550419000284.

REASON 2

American Chemical Society. n.d. "Joseph Priestley and the Discovery of Oxygen." Last modified January 20, 2022. https://www.acs.org/content/acs/en/education/whatischemistry/landmarks/josephpriestleyoxygen.html.

Bjelakovic, Goran, Dimitrinka Nikolova, Lise Lotte Gluud, Rosa G. Simonetti, and Christian Gluud. "Antioxidant Supplements for Prevention of Mortality in Healthy Participants and Patients with Various Diseases." *Cochrane Database of Systematic Reviews*, no. 3 (2012): CD007176. https://doi.org/10.1002/14651858.CD007176.pub2.

Gray, Michael W., Gertraud Burger, and B. Franz Lang. "The Origin and Early Evolution of Mitochondria." *Genome Biology* 2, no. 6 (2001): REVIEWS1018. https://doi.org/10.1186/gb-2001-2-6-reviews1018.

Harrison, Jon F., Alexander Kaiser, and John M. VandenBrooks. 2010. "Atmospheric Oxygen Level and the Evolution of Insect Body Size." *Proceedings of the Royal Society B: Biological Sciences* 277, no. 1690: 1937 –1946. https://doi.org/10.1098/rspb.2010.0001.

Pham-Huy, Lien Ai, Hua He, and Chuong Pham-Huy. "Free Radicals, Antioxidants in Disease and Health." *International Journal of Biomedical Science* 4, no. 2 (2008): 89–96. https://www.ncbi.nlm.nih.gov/pmc/articles /PMC3614697/.

Serafini, Mauro. 2006. "The Role of Antioxidants in Disease Prevention." *Medicine* 34, no. 12: 533–535. https://doi.org/10.1053/j.mpmed.2006.09.007.

Stephens, Tim. "Reign of the Giant Insects Ended with the Evolution of Birds." June 4, 2021. https://news.ucsc.edu/2012/06/giant-insects.html.

REASON 3

Enquist, Brian J., Xiao Feng, Brad Boyle, Brian Maitner, Erica A. Newman, Peter Møller Jørgensen, Patrick R. Roehrdanz, et al. "The Commonness of Rarity: Global and Future Distribution of Rarity across Land Plants." *Science Advances* 5, no. 11 (2019): eaaz0414. https://doi.org/10.1126 /sciadv.aaz0414.

Intergovernmental Panel on Climate Change. n.d. "Authors." Accessed September 27, 2022. Last modified November 7, 2022. https://www.ipcc .ch/report/ar6/wg2/about/authors/.

Intergovernmental Panel on Climate Change. "Climate Change 2022: Impacts, Adaptation and Vulnerability." Last modified November 7, 2022. https:// www.ipcc.ch/report/ar6/wg2/.

Logan, David C. "Known Knowns, Known Unknowns, Unknown Unknowns and the Propagation of Scientific Enquiry." *Journal of Experimental Botany* 60, no. 3 (2009): 712–714. https://doi.org/10.1093/jxb/erp043.

Newburger, Emma. "Climate Change Could Cost U.S. $2 Trillion Each Year by the End of the Century, White House Says." CNBC. April 4, 2022. https:// www.cnbc.com/2022/04/04/climate-change-could-cost-us-2-trillion -each-year-by-2100-omb.html.

Schultz, Emily L., Lisa Hülsmann, Michiel D. Pillet, Florian Hartig, David D. Breshears, Sydne Record, John D. Shaw, et al. "Climate-Driven, but Dynamic and Complex? A Reconciliation of Competing Hypotheses for Species' Distributions." *Ecology Letters* 25 (2022): 38–51. https://doi.org/10.1111/ele.13902.

REASON 4

Bradshaw, Corey J. A. "Little Left to Lose: Deforestation and Forest Degradation in Australia Since European Colonization," *Journal of Plant Ecology* 5, no. 1 (2012): 109–120. https://doi.org/10.1093/jpe/rtr038.

Eddy, Tyler D., Vicky W. Y. Lam, Gabriel Reygondeau, Andrés M. Cisneros-Montemayor, Krista Greer, Maria Lourdes D. Palomares, John F. Bruno, et al. "Global Decline in Capacity of Coral Reefs to Provide Ecosystem Services." *One Earth* 4, no. 9 (2021): 1278–1285. https://doi.org/10.1016/j.oneear.2021.08.016.

Farquhar, Brodie. 2021. "Wolf Reintroduction Changes Ecosystem in Yellowstone." Yellowstone National Park Trips. June 30, 2021. https://www.yellowstonepark.com/things-to-do/wildlife/wolf-reintroduction-changes-ecosystem/.

Feng, Song, and Q. Fu. "Expansion of Global Drylands under a Warming Climate." *Atmospheric Chemistry and Physics* 13, no. 6 (2013): 14637–14665. https://doi.org/10.5194/acpd-13-14637-2013.

Gauthier, Sylvie, Pierre Bernier, Timo Kuuluvainen, Anatoly Z. Shvidenko, and Dmitry G. Shchepaschenko. "Boreal Forest Health and Global Change." *Science* 349, no. 6250(2015): 819–822. https://doi.org/10.1126/science.aaa9092.

Huang, Jianping, Haipeng Yu, Xiadon Guan, Guoyin Wang, and Ruixia Guo. "Accelerated Dryland Expansion under Climate Change." *Nature Climate Change* 6 (2016): 166–171. https://doi.org/10.1038/nclimate2837.

Jones, Holly P., Peter C. Jones, Edward B. Barbier, Ryan C. Blackburn, Jose M. Rey Benayas, Karen D. Holl, Michelle McCrackin, et al. "Restoration and Repair of Earth's Damaged Ecosystems." *Royal Society Proceedings in Biological Science* 285, no. 1873 (2018): 20172577. https://doi.org/10.1098/rspb.2017.2577.

Krogh, Anders. n.d. "State of the Tropical Rainforest." Last modified May 5, 2021. https://d5i6is0eze552.cloudfront.net/documents/Publikasjoner/Andre-rapporter/RF_StateOfTheRainforest_2020.pdf?mtime=20210505115205.

Mayor, Ángeles G., Sonia Kéfi, Susana Bautista, Francisco Rodríguez, Fabrizio Cartení, and Max Reitkerk. "Feedbacks between Vegetation Pattern and Resource Loss Dramatically Decrease Ecosystem Resilience and

Restoration Potential in a Simple Dryland Model." *Landscape Ecology* 28 (2013): 931–942. https://doi.org/10.1007/s10980-013-9870-4.

Roberts, Caleb P., Dirac Twidwell, David G. Angeler, and Craig R. Allen. "How Do Ecological Resilience Metrics Relate to Community Stability and Collapse?" *Ecological Indicators* 107 (2019): 105552. https://doi.org/10.1016/j.ecolind.2019.105552.

REASON 5

Gray, James L., Leslie K. Kanagy, Edward T. Furlong, Chris J. Kanagy, Jeff W. McCoy, Andrew Mason, and Gunnar Lauenstein. "Presence of the Corexit Component Dioctyl Sodium Sulfosuccinate in Gulf of Mexico Waters after the 2010 Deepwater Horizon Oil Spill." *Chemosphere* 95 (2014): 124–130. https://doi.org/10.1016/j.chemosphere.2013.08.049.

Gyo Lee, Yong, Xavier Garza-Gomez, and Rose M. Lee. "Ultimate Costs of the Disaster: Seven Years after the Deepwater Horizon Oil Spill." *Journal of Corporate Accounting & Finance* 29, no. 1 (2018): 69–79. https://doi.org/10.1002/jcaf.22306.

Ishihara, Akiko. "Conflict Transformation Practice for Fukushima: The Past Encounters the Future through a Transformative Tour to Minamata." In *Philosophy and Practice of Bioethics Across and Between Cultures*, edited by Takao Takahashi, Nader Ghotbi, and Darryl R.J. Macer, 11. Christchurch, NZ: Eubios Ethics Institute, 2019.

Jha, Prabhat, Mary MacLennan, Frank J. Chaloupka, Ayda Yurekli, Chintanie Ramasundarahettige, Krishna Palipudi, Witold Zatońksi, et al. "Global Hazards of Tobacco and the Benefits of Smoking Cessation and Tobacco Taxes." In *Cancer: Disease Control Priorities*, 3rd ed., edited by Dean T. Jamison, Rachel Nugent, Hellen Gelband, Susan Horton, Prabhat Jha, and Ramanan Laxminarayan. Washington, DC: World Bank Publications, 2015.

Kuehn, Bridget M. "WHO: More than 7 Million Air Pollution Deaths Each Year." *JAMA* 311, no. 15 (2014): 14861486. https://doi.org/10.1001/jama.2014.4031.

Levy, Jason K., and Chennat Gopalakrishnan. "Promoting Ecological Sustainability and Community Resilience in the U.S. Gulf Coast after the 2010 Deepwater Horizon Oil Spill." *Journal of Natural Resources Policy Research* 2, no. 3 (2010): 297–315. https://doi.org/10.1080/19390459.2010.500462.

Macallister, Terry. "BP Executives Awarded Bonuses despite Deepwater Horizon Disaster." *The Guardian*. March 3, 2011. https://www.theguardian.com/business/2011/mar/03/bp-executives-bonuses-deepwater-horizon.

Margaritis, Efstathios, and Jian Kang. "Relationship between Green Space-Related Morphology and Noise Pollution." *Ecological Indicators* 72 (2017): 921–933. https://doi.org/10.1016/j.ecolind.2016.09.032.

Miller, Rossina. *Orbital Debris Quarterly News* 25, no. 4 (2021): 1–10. https://ntrs.nasa.gov/citations/20210025555.

Mimura, Nobuo, Kazuya Yasuhara, Seiki Kawagoe, Hiromune Yokoki, and So Kazama. "Damage from the Great East Japan Earthquake and Tsunami—A Quick Report." *Mitigation and Adaptation Strategies for Global Change* 16 (2011): 803–818. https://doi.org/10.1007/s11027-011-9297-7.

Orru, Hans, Kristie Ebi, and Bertil Forsberg. "The Interplay of Climate Change and Air Pollution on Health." *Current Environmental Health Reports* 4, no. 5 (2017): 504–513. https://doi.org/10.1007/s40572-017-0168-6.

Papastefanou, Constantin. "Escaping Radioactivity from CoalFired Power Plants (CPPs) Due to Coal Burning and the Associated Hazards: A Review." *Journal of Environmental Radioactivity* 101, no. 3 (2010): 191–200. https://doi.org/10.1016/j.jenvrad.2009.11.006.

Papp, Z., Z. Dezsö, and S. Daroczy. "Significant Radioactive Contamination of Soil around a Coal-Fired Thermal Power Plant." *Journal of Environmental Radioactivity* 59, no. 2 (2002): 191–205. https://doi.org/10.1016/S0265-931X(01)00071-6.

Siqueira, Diana Silva, Josué de Almeida Meystre, Maicon Quieroz Hilário, Danilo Henrique Donato Rocha, Genésio José Menon, and Rogério José da Silva. "Current Perspectives on Nuclear Energy as a Global Climate Change Mitigation Option." *Mitigation and Adaptation Strategies for Global Change* 24 (2019): 749–777. https://doi.org/10.1007/s11027-018-9829-5.

U.S. Census Bureau. n.d. "Real Median Personal Income in the United States." FRED Economic Data. Last modified September 13, 2022. https://fred.stlouisfed.org/series/MEPAINUSA672N.

Wong, Kaufui V., Andrew Paddon, and Alfredo Jimenez. "Review of World Urban Heat Islands: Many Linked to Increased Mortality." *Journal of Energy Resources Technology* 135, no. 2 (2013): 022101–022112. https://doi.org/10.1115/1.4023176.

REASON 6

Cascio, Jessica, and E. Ashby Plant. "Prospective Moral Licensing: Does Anticipating Doing Good Later Allow You to Be Bad Now?" *Journal of Experimental Social Psychology* 56 (2015): 110–116. https://doi.org/10.1016/j.jesp.2014.09.009.

Darwin, Charles, and Leonard Kebler. *On the Origin of Species by Means of Natural Selection, or The Preservation of Favored Races in the Struggle for Life*. London: J. Murray, 1859.

Ghosh, Pallab. 2020. "Sex Robots May Cause Psychological Damage." BBC News. February 15, 2020. https://www.bbc.com/news/science-environment-51330261.

Kaas, Jon H. "The Evolution of Brains from Early Mammals to Humans." *Wiley Interdisciplinary Reviews: Cognitive Science* 4, no. 1 (2012): 33–45. https://doi.org/10.1002/wcs.1206.

Krockow, Eva M. "Is a Zero-Risk Bias Impairing Your Crisis Response?" *Psychology Today*. March 19, 2020. https://www.psychologytoday.com/au/blog/stretching-theory/202003/is-zero-risk-bias-impairing-your-crisis-response.

Merritt, Anna, Daniel Effron, and Benoît Monin. "Moral Self-Licensing: When Being Good Frees Us to Be Bad." *Social and Personality Psychology Compass* 4, no. 5 (2010): 344–357. https://doi.org/10.1111/j.1751-9004.2010.00263.x.

REASON 7

Drexler, K. Eric. *Engines of Creation*. New York: Anchor Books, 1986.

Ogas, Ogi, and Sai Gaddam. *A Billion Wicked Thoughts: What the Internet Tells Us About Sexual Relationships*. New York: Penguin Books, 2011.

Thompson, Avery. "Scientists Have Made Transistors Smaller Than We Thought Possible." *Popular Mechanics*. October 12, 2016. https://www.popularmechanics.com/technology/a23353/1nm-transistor-gate/.

Zafar, Ramish. "Apple A13 for iPhone 11 Has 8.5 Billion Transistors, Quad-Core GPU." Wccftech. September 10, 2019. https://wccftech.com/apple-a13-iphone-11-transistors-gpu/.

REASON 8

BBC News. "Hiroshima and Nagasaki: 75th Anniversary of Atomic Bombings." August 9, 2020. https://www.bbc.com/news/in-pictures-53648572.

Brode, Harold L. "Review of Nuclear Weapons Effects." *Annual Review of Nuclear Science* 18, no. 1 (1986): 153–202. https://doi.org/10.1146/annurev.ns.18.120168.001101.

Kolb, W. M. and P. G. Carlock. n.d. "Trinitite." ORAU: Museum of Radiation and Radioactivity. Last modified April 18, 2022. https://www.orau.org/health-physics-museum/collection/nuclear-weapons/trinity/trinitite.html.

Kurzgesagt—In a Nutshell. "What If We Detonated All Nuclear Bombs at Once?" YouTube. March 31, 2019.https://youtu.be/JyECrGp-Sw8.

National Archives. "The Atomic Bombing of Hiroshima and Nagasaki, August 1945." Last modified October 25, 2022. https://www.archives.gov/news /topics/hiroshima-nagasaki-75.

Office of Legacy Management. n.d. "Trinity Site—World's First Nuclear Explosion." U.S. Department of Energy. Last modified December 29, 2021. https://www.energy.gov/lm/doe-history/manhattan-project-background -information-and-preservation-work/manhattan-project-1.

O'Hagan, Sean. "Armed to the Milk Teeth: America's GunToting Kids." *The Guardian*. April 29, 2014. https://www.theguardian.com/artanddesign/2014 /apr/29/armed-to-the-milk-teeth-america-gun-toting-kids.

PlenilunePictures. "J. Robert Oppenheimer: 'I Am Become Death, the Destroyer of Worlds.'" YouTube. August 6, 2021. https://youtu.be/lb13ynu3Iac.

Reisner, Jon, Gennaro D'Angelo, Eunmo Koo, Wesley Even, Matthew Hecht, Elizabeth Hunke, Darin Comeau, et al. "Climate Impact of a Regional Nuclear Weapons Exchange: An Improved Assessment Based on Detailed Source Calculations." *Journal of Geophysical Research: Atmospheres* 123, no. 5 (2018): 2752–2772. https://doi.org/10.1002/2017jd027331.

Robock, Alan, and Owen Brian Toon. "Self-assured Destruction: The Climate Impacts of Nuclear War." *Bulletin of the Atomic Scientists* 68, no. 5 (2015): 66–74. https://doi.org/10.1177/0096340212459127.

Roser, Max, Bastian Herre, and Joe Hasell. n.d. "Nuclear Weapons." Our World in Data. Last modified September 18, 2022. https://our worldindata.org/nuclear-weapons.

Toon, Owen Brian, Alan Robock, and Rich P. Turco. "Environmental Consequences of Nuclear War." *Physics Today* 61, no. 12 (2008): 37–42. https://doi.org/10.1063/1.3047679.

World Population Review. "Nuclear Weapons by Country 2022." Last modified October 23, 2022. https://worldpopulationreview.com/country-rankings /nuclear-weapons-by-country.

REASON 9

Foster, Timothy J. "Antibiotic Resistance in *Staphylococcus aureus*." *FEMS Microbiology Reviews* 41, no. 3 (2017): 430–449. https://doi.org/10.1093 /femsre/fux007.

Grand View Research. "Alcoholic Drinks Market Size, Share and Trends Analysis Report by Type (Beer, Spirits, Wine, Cider, Perry and Rice

Wine, Hard Seltzer), by Distribution Channel, by Region, and Segment Forecasts, 2022–2028." Last modified June 4, 2022. https://www.grand viewresearch.com/industry-analysis/alcoholic-drinks-market-report.

IPCC. *Climate Change 2022: Synthesis Report. Contribution of Working Groups I, II and III to the Sixth Assessment Report of the Intergovernmental Panel on Climate Change.* Geneva, Switzerland: IPCC, 2022.

Llor, Carl, and Lars Bjerrum. "Antimicrobial Resistance: Risk Associated with Antibiotic Overuse and Initiatives to Reduce the Problem." *Therapeutic Advances in Drug Safety* 5, no. 6 (2014): 229–241. https://doi.org/10.1177/2042098614554919.

Lowy, Franklin D. "Antimicrobial Resistance: The Example of *Staphylococcus aureus.*" *Journal of Clinical Investigation* 111, no. 9 (2003): 1265–73. https://doi.org/10.1172/JCI18535.

Manyi-Loh, Christy, Sampson Mamphweli, Edson Meyer, and Anthony Okoh. "Antibiotic Use in Agriculture and Its Consequential Resistance in Environmental Sources: Potential Public Health Implications." *Molecules* 23, no. 4 (2018): 795. https://doi.org/10.3390/molecules23040795.

Omulo, Sylvia, Samuel M. Thumbi, M. Kariuki Njenga, and Douglas R. Call. "A Review of 40 Years of Enteric Antimicrobial Resistance Research in Eastern Africa: What Can Be Done Better?" *Antimicrobial Resistance and Infection Contro.* 4, no. 1 (2015). https://doi.org/10.1186/s13756-014-0041-4.

Taveira, Iasmin Cartaxo, Karoline Maria Vieira Nogueira, Débora Lemos Gadelha De Oliveira, and Roberto Do Nascimento Silva. "Fermentation: Humanity's Oldest Biotechnological Tool." *Frontiers for Young Minds* 9 (2021): 568656. https://doi.org/10.3389/frym.2021.568656.

REASON 10

Bostrom, Nick. "Ethical Issues in Advanced Artificial Intelligence." Last modified June 1, 2022. https://nickbostrom.com/ethics/ai.

Bostrom, Nick. *Superintelligence: Paths, Dangers, Strategies.* Oxford: Oxford University Press.

Hern, Alex. "Experts Including Elon Musk Call for Research to Avoid AI 'Pitfalls.'" *The Guardian.* January 12, 2015. https://www.theguardian.com /technology/2015/jan/12/elon-musk-ai-artificial-intelligence-pitfalls.

NOVA. "What's the Next Big Thing?" Last modified May 13, 2022. https://www.pbs.org/wgbh/nova/video/whats-the-next-big-thing/.

Schwartz, Oscar. "In 2016, Microsoft's Racist Chatbot Revealed the Dangers of Online Conversation." IEEE Spectrum. November 25, 2019. https://

spectrum.ieee.org/in-2016-microsofts-racist-chatbot-revealed-the
-dangers-of-online-conversation.

Turing, Alan M. "I.—Computing Machinery and Intelligence." *Mind* 59, no. 236 (195): 433–460. https://doi.org/10.1093/mind/LIX.236.433.

REASON 11

Blanchett, Amy, and Laurie Abadie. "Space Radiation Is Risky Business for the Human Body." NASA. Last modified August 16, 2018. https://www.nasa .gov/feature/space-radiation-is-risky-business-for-the-human-body.

Chakraborty, Ranjani, and Melissa Hirsch. 2020. "The Biggest Radioactive Spill in U.S. History Never Ended." Vox, October 13, 2020. https://www.vox .com/21514587/navajo-nation-new-mexico-radioactive-uranium-spill.

Cucinotta, Francis A., and Marco Durante. "Cancer Risk from Exposure to Galactic Cosmic Rays: Implications for Space Exploration by Human Beings." *Lancet Oncology* 7, no. 5 (2006): 431–435. https://doi.org/10.1016 /S1470-2045(06)70695-7.

Goldsmith, R., J. D. Boice Jr., Z. Hrubec, P. E. Hurwitz, T. E. Goff, and J. Wilson. "Mortality and Career Radiation Doses for Workers at a Commercial Nuclear Power Plant: Feasibility Study." *Health Physics* 56, no. 2 (1989): 139–50. https://doi.org/10.1097/00004032-198902000-00001.

ICRP. "The 2007 Recommendations of the International Commission on Radiological Protection." ICRP Publication 103. *Annals of the ICRP* 37 (2007): 2–4.

Hahn, Trish. "I Fell on It: Tales from the Dirt Track." Interview by Byrne LaGinestra and Wade Fariclough. *Sci-gasm.* Podcast. April 18, 2019. https://omny.fm/shows/sci-gasm/i-fell-on-it-tales-from-the-dirt-track.

Lee, Stan, and Jack Kirby. *The Incredible Hulk.* New York: Marvel Comics, 1963.

Lundin, Frank E. Jr., William J. Lloyd, and Elizabeth M. Smith. "Mortality of Uranium Miners in Relation to Radiation Exposure, Hard-Rock Mining and Cigarette Smoking—1950 through September 1967." *Health Physics* 16, no. 5 (1969): 571–578. https://doi.org/10.1097/00004032-196905000-00004.

Matanoski G.M., R. Seltser, P. E. Sartwell, E. L. Diamond, and E. A. Elliott. "The Current Mortality Rates of Radiologists and Other Physician Specialists: Specific Causes of Death." *American Journal of Epidemiology* 101, no. 3 (1975): 199–210. https://doi.org/10.1093/oxfordjournals.aje.a112087.

Matthews, Natalie H., Wen-Qing Li, Abrar A. Qureshi, Martin A. Weinstock, and Eunyoung Choo. "Epidemiology of Melanoma." In *Cutaneous*

Melanoma: Etiology and Therapy, edited by William H. Ward and Jeffrey M. Farma. Brisbane: Codon Publications, 2017. https://doi.org/10.15586 /codon.cutaneousmelanoma.2017.ch1.

Mohler, Stanley R. "Galactic Radiation Exposure during Commercial Flights: Is There a Risk?" *Canadian Medical Association Journal* 168, no. 9 (2003): 1157–1158.

Moore, Kate. *The Radium Girls: The Dark Story of America's Shining Women.* Naperville, IL: Sourcebooks, 2017.

Papastefanou, C. "Radiation Dose from Cigarette Tobacco." *Radiation Protection Dosimetry* 123, no. 1 (2007): 68–73. https://doi.org/10.1093/rpd/ncl033.

United States Nuclear Regulatory Commission. "High Radiation Doses." Last modified March 20, 2020. https://www.nrc.gov/about-nrc/radiation /health-effects/high-rad-doses.html.

REASON 12

Abumrad, Nada A. "CD36 May Determine Our Desire for Dietary Fats." *Journal of Clinical Investigation* 115, no. 11 (2005): 2965–2967. https:// doi.org/10.1172/jci26955.

Forouhi, Nita G., and Nigel Unwin. "Global Diet and Health: Old Questions, Fresh Evidence, and New Horizons." *The Lancet* 393, no. 10184 (2019): 1916–1918. https://doi.org/10.1016/S0140-6736(19)30500-8.

GBD 2017 Diet Collaborators. "Health Effects of Dietary Risks in 195 Countries, 1990–2017: A Systematic Analysis for the Global Burden of Disease Study 2017." *The Lancet* 393, no. 10184 (2019): 1958–1972, https:// doi.org/10.1016/S0140-6736(19)30041-8.

Hewings-Martin, Yella. "This Is How Taste Keeps Us Safe." Medical News Today. August 10, 2017. https://www.medicalnewstoday.com /articles/318874.

Keller, Kathleen L., Lisa C. H. Liang, Johannah Sakimura, Daniel May, Christopher van Belle, Cameron Breen, Elissa Driggin, et al. "Common Variants in the CD36 Gene Are Associated with Oral Fat Perception, Fat Preferences, and Obesity in African Americans." *Obesity* 20, no. 5 (2012): 1066–1073. https://doi.org/10.1038/oby.2011.374.

Li, Feng. "Taste Perception: From the Tongue to the Testis." *Molecular Human Reproduction* 19, no. 6 (2013): 349–360. https://doi.org/10.1093/molehr /gat009.

Moffat, Tina, and Shanti Morell-Hart. "How the Mediterranean Diet Became No. 1—and Why That's a Problem." The Conversation. February 24, 2020.

https://theconversation.com/how-the-mediterranean-diet-became-no
-1-and-why-thats-a-problem-131771.

Roxby, Philippa, 2014. "What Leonardo Taught Us about the Heart." BBC
News. June 28, 2014. https://www.bbc.com/news/health-28054468.

Running, Cordelia A., Bruce A. Craig, and Richard D. Mattes. "Oleogustus:
The Unique Taste of Fat." *Chemical Senses* 40, no. 7 (2015): 507–516.
https://doi.org/10.1093/chemse/bjv036.

REASON 13

Goronzy, Jörg J., and Cornelia M. Weyand. "Immune Aging and Autoimmunity."
Cellular and Molecular Life Sciences 69 (2012): 1615–1623. https://doi.
org/10.1007/s00018-012-0970-0.

Hayter, Scott M., and Matthew C. Cook. "Updated Assessment of the
Prevalence, Spectrum and Case Definition of Autoimmune Disease."
Autoimmunity Reviews 11, no. 10 (2012): 754–765. https://doi.org/10.1016/j.
autrev.2012.02.001.

Hepworth, Matthew R., Laurel A. Monticelli, Thomas C. Fung, Carly G. K. Ziegler,
Stephanie Grunberg, Rohini Sinha, Adriena R. Mantegazza, et al. "Innate
Lymphoid Cells Regulate CD4 T-cell Responses to Intestinal Commensal
Bacteria." *Nature* 498 (2013): 113–117. https://doi.org/10.1038/nature12240.

Orbai, Ana-Maria. "What Are Common Symptoms of Autoimmune
Disease?" John Hopkins Medicine. Last modified September 26, 2022.
https://www.hopkinsmedicine.org/health/wellness-and-prevention
/what-are-common-symptoms-of-autoimmune-disease.

Watson, Stephanie. "Autoimmune Diseases: Types, Symptoms, Causes, and
More." Healthline. Last modified July 15, 2022. https://www.healthline
.com/health/autoimmune-disorders.

REASON 14

Aubert, Geraldine, and Peter M. Lansdorp. "Telomeres and Aging."
Physiological Reviews 88, no. 2 (2008): 557–79. https://doi.org/10.1152
/physrev.00026.2007.

Bartels, Meike. "Genetics of Wellbeing and Its Components Satisfaction
with Life, Happiness, and Quality of Life: A Review and Meta-analysis
of Heritability Studies." *Behavior Genetics* 45, no. 2: 137–156. https://
doi.org/10.1007/s10519-015-9713-y.

Chetty, Raj, Michael Stepner, Sarah Abraham, Shelby Lin, Benjamin Scuderi,
Nicholas Turner, Augustin Bergeron, et al. "The Association Between

Income and Life Expectancy in the United States, 2001–2014." *JAMA* 315, no. 16 (2016):1750–66. https://doi.org/10.1001/jama.2016.4226.

Edenberg, Howard J., and Tatiana Foroud. "Genetics and Alcoholism." *Nature Reviews Gastroenterology & Hepatology* 10, no. 8 (2013): 487–494. https://doi.org/10.1038/nrgastro.2013.86.

Fan, Yuxin, Elena Linardopoulou, Cynthia Friedman, Eleanor Williams, and Barbara J. Trask. "Genomic Structure and Evolution of the Ancestral Chromosome Fusion Site in 2q13–2q14.1 and Paralogous Regions on Other Human Chromosomes." *Genome Research* 12, no. 11 (2002): 1651–1662, https://doi.org/10.1101/gr.337602.

Haldeman-Englert, Chad, Donna Freeborn, and Raymond Kent Turley. n.d. "Autosomal Recessive: Cystic Fibrosis, Sickle Cell Anemia, Tay Sachs Disease—Health Encyclopedia." University of Rochester Medical Center. Last modified July 1, 2022. https://www.urmc.rochester.edu/encyclopedia/content.aspx?ContentID=P02142&ContentTypeID=90#:~:text=Sickle%20cell%20anemia%20is%20another.

Learn.Genetics. n.d. "Are Telomeres the Key to Aging and Cancer." Last modified October 9, 2022. https://learn.genetics.utah.edu/content/basics/telomeres/#:~:text=Cawthon.

National Library of Medicine—Profiles in Science. "The Discovery of the Double Helix, 1951–1953." Last modified February 14, 2022. https://profiles.nlm.nih.gov/spotlight/sc/feature/doublehelix.

REASON 15

Bruce-Keller, Annadora J., J. Michael Salbaum, and Hans-Rudolf Berthoud. "Harnessing Gut Microbes for Mental Health: Getting from Here to There." *Biological Psychiatry* 83, no. 3 (2018): 214–223. https://doi.org/10.1016/j.biopsych.2017.08.014.

Kaneda, Toshiko, and Carl Haub. "How Many People Have Ever Lived on Earth?" Population Reference Bureau. Last modified November 8, 2022. https://www.prb.org/articles/how-many-people-have-ever-lived-on-earth/.

Schloss, Patrick D., and Jo Handelsman. "Status of the Microbial Census." *Microbiology and Molecular Biology Reviews* 68, no. 4 (2004): 686–91. https://doi.org/10.1128/MMBR.68.4.686–691.2004.

Sender, Ron, Shai Fuchs, and Ron Milo. "Revised Estimates for the Number of Human and Bacteria Cells in the Body." *PLOS Biology* 14, no. 8 (2016): 1–14. https://doi.org/10.1371/journal.pbio.1002533.

Spyrou, Maria A., Lyazzat Musralina, Guido A. Gnecchi Ruscone, Arthur

Kocher, Pier-Giorgio Borbone, Valeri I. Khartanovich, Alexandra Buzhilova, et al. "The Source of the Black Death in Fourteenth-Century Central Eurasia." *Nature* 606, no. 7915 (2022): 718–724. https://doi.org/10.1038/s41586-022-04800-3.

Von Wintersdorff, Christian J., John Penders, Julius M. Van Niekerk, Nathan D. Mills, Snehali Majumder, Lieke B. Van Alphen, Paul H. M. Savelkoul, et al. "Dissemination of Antimicrobial Resistance in Microbial Ecosystems through Horizontal Gene Transfer." *Frontiers in Microbiology* 7 (2016): 173. https://doi.org/10.3389/fmicb.2016.00173.

REASON 16

Amazonas, Diana R., José A. Portes-Junior, Milton Y. Nishiyama-Jr, Carolina A. Nicolau, Hipócrates M. Chalkidis, Rosa H. V. Mourão, Felipe G. Grazziotin, et al. "Molecular Mechanisms Underlying Intraspecific Variation in Snake Venom." *Journal of Proteomics* 181 (2018): 60–72. https://doi.org/10.1016/j.jprot.2018.03.032.

Barazzone, Constance, Stuart Horowitz, Yves R. Donati, Ivan Rodriguez, and Pierre-François Piguet. "Oxygen Toxicity in Mouse Lung: Pathways to Cell Death." *American Journal of Respiratory Cell and Molecular Biology* 19, no. 4 (1998): 573–581. https://doi.org/10.1165/ajrcmb.19.4.3173.

Bloch, Harry. "Poisons and Poisoning: Implication of Physicians with Man and Nations." *Journal of the National Medical Association* 79, no. 7 (1987): 761–764. https://www.ncbi.nlm.nih.gov/pmc/articles/PMC2625561/pdf/jnma00922-0097.pdf.

Centers for Disease Control and Prevention. n.d. "Facts about Cyanide." Last modified April 4, 2018. https://emergency.cdc.gov/agent/cyanide/basics/facts.asp.

Clarke, Suzan, and Rich McHugh. "Jury Rules against Radio Station after Water-Drinking Contest Kills Calif. Mom." ABC News. November 2, 2009. https://abcnews.go.com/GMA/jury-rules-radio-station-jennifer-strange-water-drinking/story?id=8970712.

Cotton, Simon. "Handle with Care—The World's Five Deadliest Poisons." The Conversation. April 12, 2016. https://theconversation.com/handle-with-care-the-worlds-five-deadliest-poisons-56089.

Mach, William J., Amanda R. Thimmesch, J. Thomas Pierce, and Janet D. Pierce. "Consequences of Hyperoxia and the Toxicity of Oxygen in the Lung." *Nursing Research and Practice* 2011. https://doi.org/10.1155/2011/260482.

REASON 17

Cooper, Geoffrey M. "The Origin and Evolution of Cells." In *The Cell: A Molecular Approach*, 2nd ed. Sunderland, MA: Sinauer Associates, 2000. https://www.ncbi.nlm.nih.gov/books/NBK9841/.

Englar, Ryane E. "Spines along the Feline Penis." In *Common Clinical Presentations in Dogs and Cats*, edited by Ryane E. Englar. New York: John Wiley & Sons, 2019. https://doi.org/10.1002/9781119414612.

Fisher, Diana O., Christopher R. Dickman, Menna E. Jones, and Simon P. Blomberg. "Sperm Competition Drives the Evolution of Suicidal Reproduction in Mammals." *Proceedings of the National Academy of Sciences* 110, no. 44 (2013): 17910–17914. https://doi.org/10.1073/pnas.1310691110.

Girard, Madeline B., Damian O. Elias, and Michael M. Kasumovic. "Female Preference for Multi-modal Courtship: Multiple Signals Are Important for Male Mating Success in Peacock Spiders." *Proceedings of the Royal Society B: Biological Sciences* 282, no. 1820 (2015): 20152222. https://doi.org/10.1098/rspb.2015.2222.

Kristan, William B. "Early Evolution of Neurons." *Current Biology* 26, no. 20 (2016): R949–R954. https://doi.org/10.1016/j.cub.2016.05.030.

Morrow, Edward H., and Göran Arnqvist. "Costly Traumatic Insemination and a Female Counter-adaptation in Bed Bugs." *Proceedings of the Royal Society of London. Series B: Biological Sciences* 270, no. 1531 (2003): 2377–2381. https://doi.org/10.1098/rspb.2003.2514.

Naylor, Ryan, Samantha J. Richardson, and Bronwyn M. McAllan. "Boom and Bust: A Review of the Physiology of the Marsupial Genus *Antechinus*." *Journal of Comparative Physiology B* 178, no. 5 (2008): 545–562. https://doi.org/10.1007/s00360-007-0250-8.

REASON 18

Eckhardt, William. "Probability Theory and the Doomsday Argument." *Mind* 102, no. 407 (1993): 483–488. https://doi.org/10.1093/mind/102.407.483.

Gott, J. Richard. "Implications of the Copernican Principle for Our Future Prospects." *Nature* 363: 315–319. https://doi.org/10.1038/363315a0.

Richmond, Alasdair. "The Doomsday Argument." *Philosophical Books* 47, no. 2 (2006): 129–142. https://doi.org/10.1111/j.1468-0149.2006.00392.x.

United Nations. "World Population Prospects 2022." United Nations. Last modified September 29, 2022. https://population.un.org/wpp/.

REASON 19

Australian Institute of Health and Welfare. 2021. "Venomous Bites and Stings, 2017–18." Australian Government. Last modified March 9, 2021. https://www.aihw.gov.au/getmedia/78b416bf-0250-4368-89d6-81e2d9f32528/aihw-injcat-215.pdf.aspx.

Berns, Gregory S., Andrew M. Brooks, and Mark Spivak. "Functional MRI in Awake Unrestrained Dogs." *PLOS One* 7, no. 5 (2012): e38027. https://doi.org/10.1371/journal.pone.0038027.

Centers for Disease Control and Prevention. "Nonfatal Dog Bite–Related Injuries Treated in Hospital Emergency Departments—United States, 2001." *Morbidity and Mortality Weekly Report* 52, no. 26 (2003): 605–610. https://www.cdc.gov/mmwr/preview/mmwrhtml/mm5226a1.htm.

Growth from Knowledge. "Man's Best Friend: Global Pet Ownership and Feeding Trends." November 11, 2016, last modified May 6, 2022. https://www.gfk.com/insights/mans-best-friend-global-pet-ownership-and-feeding-trends.

Hui Gan, Genieve Z., Anne-Marie Hill, Polly Yeung, Sharon Keesing, and Julie A. Netto. "Pet Ownership and Its Influence on Mental Health in Older Adults." *Aging & Mental Health* 24, no. 10 (2020): 1605–1612. https://doi.org/10.1080/13607863.2019.1633620.

Johnson, Murray. "'Feathered Foes': Soldier Settlers and Western Australia's 'Emu War' of 1932." *Journal of Australian Studies* 30, no. 88 (2006): 147–157. https://doi.org/10.1080/14443050609388083.

Jones, Kate E., Nikkita G. Patel, Marc A. Levy, Adam Storeygard, Deborah Balk, John L. Gittleman, and Peter Daszak. "Global Trends in Emerging Infectious Diseases." *Nature* 451, no. 7181 (2008): 990–993. https://doi.org/10.1038/nature06536.

Karl, Sabrina, Magdalena Boch, Anna Zamansky, Dirk van der Linden, Isabella C. Wagner, Christoph J. Völter, Claus Lamm, et al. "Exploring the Dog-Human Relationship by Combining fMRI, Eye-Tracking and Behavioral Measures." *Scientific Reports* 10, no. 1 (2003): 1–15. https://doi.org/10.1038/s41598-020-79247-5.

Mubanga, Mwenya, Liisa Byberg, Agneta Egenvall, Erik Ingelsson, and Tove Fall. "Dog Ownership and Survival after a Major Cardiovascular Event: A Register-Based Prospective Study." *Circulation: Cardiovascular Quality and Outcomes* 12, no. 10 (2019): e005342. https://doi.org/10.1161/CIRCOUTCOMES.118.005342.

National Institutes of Health. "Infectious Disease Emergence: Past, Present, and Future." In *Microbial Evolution and Co-Adaptation: A Tribute to the Life and*

Scientific Legacies of Joshua Lederberg: Workshop Summary. Washington, D.C.: National Academies Press, 2009. https://www.ncbi.nlm.nih.gov /books/NBK45714/.

Newgate Research. "Pets in Australia: A National Survey of Pets and People." Animal Medicines Australia. Last modified October 21, 2022. https://animalmedicinesaustralia.org.au/wp-content/uploads/2019/10 /ANIM001-Pet-Survey-Report19_v1.7_WEB_high-res.pdf.

Tuckel, Peter S., and William Milczarski. "The Changing Epidemiology of Dog Bite Injuries in the United States, 2005–2018." *Injury Epidemiology* 7, no. 1 (2020): 57. https://doi.org/10.1186/s40621-020-00281-y.

Wanshel, Elyse. "Who Loves Their Humans More—Cats or Dogs? Here's the Answer." HuffPost. February 1, 2016. Last modified March 12, 2016. https://www.huffpost.com/entry/cat-vs-dog-who-loves-humans-more _n_56af85a4e4b077d4fe8ed1ed.

REASON 20

Geller, Robert J. "Earthquake Prediction: A Critical Review." *Geophysical Journal International* 131, no. 3 (1997): 425–450, https://doi.org/10.1111 /j.1365-246X.1997.tb06588.x.

Kanamori, Hiroo. "Earthquake Prediction: An Overview." *International Geophysics* 81, part B (2003): 1205–1216. https://doi.org/10.1016 /S0074-6142(03)80186-9.

REASON 21

Bye, Bente L. "Volcanic Eruptions: Science and Risk Management." Science 2.0. May 27, 2011. Last modified November 4, 2022. https:// www.science20.com/planetbye/volcanic_eruptions_science_and _risk_management-79456.

Ferracane, Jessica. 2019. "Visitation to Hawai'i Volcanoes National Park Creates $94.1 Million in Economic Benefits." National Parks Service. May 24, 2019. https://www.nps.gov/havo/learn/news/20190524_hvnpeconbenefits.htm.

Gibbons, Ann. "Pleistocene Population Explosions: A Controversial Method of Reconstructing Prehistorical Populations Indicates That Separate Modern Human Groups—and Not a Single Group from Africa—Suddenly Expanded about 50,000 Years Ago." *Science* 262, no. 5130 (1993): 27–28. https:// doi.org/10.1126/science.262.5130.27.

Green, Theodore, Paul R. Renne, and C. Brenhin Keller. "Continental Flood Basalts Drive Phanerozoic Extinctions." *Proceedings of the National Academy of Sciences* 119, no. 38 (2022): e2120441119. https://doi.org/10.1073 /pnas.2120441119.

Grisham, Lori. 2015. "'I'm Going to Stay Right Here.' Lives Lost in Mount St. Helens Eruption." USA Today. May 17, 2015. https://www.usa today.com/story/news/nation-now/2015/05/17/mount-st-helens-people -stayed/27311467/.

Mastin, Larry G., Alexa R. Van Eaton, and Jacob B. Lowenstern. "Modeling Ash Fall Distribution from a Yellowstone Supereruption." *Geochemistry, Geophysics, Geosystems* 15, no. 8 (2014): 3459–3475. https:// doi.org/10.1002/2014GC005469.

Rampino, Michael R., and Stephen Self. "Climate-Volcanism Feedback and the Toba Eruption of 74,000 Years Ago." *Quaternary Research* 40, no. 3 (1993): 269–280. https://doi.org/10.1006/qres.1993.1081.

Warthin, Morgan. "Tourism to Yellowstone Creates $560 Million in Economic Benefits." National Parks Service. June 15, 2021, https://www.nps.gov /yell/learn/news/21016.htm.

REASON 22

Chaudhary, Chhaya, and Mark J. Costello. "Species Richness Decreases with Depth in the Ocean." *Deep-Sea Life* 10 (2017): 11–12.

Costello, Mark J., and Chhaya Chaudhary. "Marine Biodiversity, Biogeography, Deep-Sea Gradients, and Conservation." *Current Biology* 27, no. 11 (2017): R511–R527. https://doi.org/10.1016/j.cub.2017.04.060.

Earle, Steven. "The Temperature of Earth's Interior." In *Physical Geography*, 2nd ed., edited by Steven Earle. BC Campus OpenEd. 2019. https:// opentextbc.ca/geology/chapter/9-2-the-temperature-of-earths-interior/.

Kidder, Stanley Q., and Thomas H. Vonder Haar. "Winds." In *Satellite Meteorology*, edited by Stanley Q. Kidder and Thomas H. Vonder Haar, 233–258. Cambridge, MA: Academic Press, 1995. https:// doi.org/10.1016/B978-0-08-057200-0.50011-0.

Newcastle University. "New Species Discovered in the Ultra Deep." Newcastle University. September 10, 2018. https://www.ncl.ac.uk/press/articles /archive/2018/09/threenewspecies/.

REASON 23

Associated Press. "Local Officials Nearly Fall for H2O Hoax." NBC News. March 16, 2004, https://www.nbcnews.com/id/wbna4534017.

Gnad, Megan. "MP Tries to Ban Water." NZ Herald. September 14, 2007.

https://www.nzherald.co.nz/nz/mp-tries-to-ban-water/XM4GJ7XG3
WC4ANBIFP2IVFNANE/?c_id=1&objectid=10463579.

New Zealand National Party. "Greens Support Ban on Water!" Scoop
Independent News. October 25, 2001. https://www.scoop.co.nz/stories
/PA0110/S00440.htm.

Treacy, Josephine. "Drinking Water Treatment and Challenges in Developing
Countries." In *The Relevance of Hygiene to Health in Developing Countries*,
edited by Natasha Potgieter and Afsatou Ndama Traore Hoffman, 55–77.
London: InTech Open, 2019. https://doi.org/10.5772/intechopen.80780.

Weeks, W. F., and W. Campbell. "Icebergs as a Fresh-Water Source: An
Appraisal." *Journal of Glaciology* 12, no. 65 (1973): 207–233. https://
doi.org/10.3189/S0022143000032044.

REASON 24

Benton, Michael J. "Palaeontological Data and Identifying Mass Extinctions."
Trends in Ecology & Evolution 9, no. 5 (1994): 181–185. https://
doi.org/10.1016/0169-5347(94)90083-3.

Berkner, L. V., and L. C. Marshall. "On the Origin and Rise of Oxygen Concentration
in the Earth's Atmosphere." *Journal of Atmospheric Sciences* 22, no. 3 (1965): 225–
261. https://doi.org/10.1175/1520–0469(1965)022<0225:OTOARO>2.0.CO;2.

Darwin, Charles, and Leonard Kebler. *On the Origin of Species by Means of
Natural Selection, or The Preservation of Favored Races in the Struggle for
Life*. London: John Murray, 1869.

Goodenough, Anne E., Natasha Little, William S. Carpenter, and Adam
G. Hart. "Birds of a Feather Flock Together: Insights into Starling
Murmuration Behavior Revealed Using Citizen Science." *PLOS One* 12,
no. 6 (2017): e0179277. https://doi.org/10.1371/journal.pone.0179277.

Hamer, Ashley. "99 Percent of the Earth's Species Are Extinct—But That's Not
the Worst of It." Discovery. August 1, 2019. https://www.discovery.com
/nature/99-Percent-Of-The-Earths-Species-Are-Extinct.

Hedges, S. Blair. "The Origin and Evolution of Model Organisms." *Nature
Reviews Genetics* 3, no. 11 (2002): 838–849. https://doi.org/10.1038/nrg929.

Poynton, Howard. "Pondering Petrichor: The Smell of Rain: How CSIRO
Invented a New Word." *Chemistry in Australia*, September 2015: 34–35.
https://search.informit.org/doi/10.3316/informit.464822231422788.

Raup, David M., and J. John Sepkoski Jr. "Mass Extinctions in the Marine
Fossil Record." *Science* 215, no. 4539 (1982): 1501–1503. https://
doi.org/10.1126/science.215.4539.1501.

REASON 25

Abe, Y. "A Proto-atmosphere and the Environment of the Earth During Accretion." *American Geophysical Union*, December 2001. https://ui.adsabs.harvard.edu/abs/2001AGUFM.U51A..09A/abstract.

Björn, Lars Olof, and Govindjee. "The Evolution of Photosynthesis and Its Environmental Impact." In *Photobiology*, edited by Lars Olof Björn. New York: Springer, 2008. https://doi.org/10.1007/978-0-387-72655-7_12.

Cardona, Tanai, and A. William Rutherford. "Evolution of Photochemical Reaction Centers: More Twists?" *Trends in Plant Science* 24, no. 11 (2019): 1008–1021. https://doi.org/10.1016/j.tplants.2019.06.016.

He, Tianchen, Maoyan Zhu, Benjamin J. W. Mills, Peter M. Wynn, Andrey Yu. Zhuravlev, Rosalie Tostevin, Philip A.E. Pogge von Strandmann, et al. "Possible Links between Extreme Oxygen Perturbations and the Cambrian Radiation of Animals." *Nature Geoscience* 12, no. 6 (2019): 468–474. https://doi.org/10.1038/s41561-019-0357-z.

Kopp, Robert E., Joseph L. Kirschvink, Isaac A. Hilburn, and Cody Z. Nash. "The Paleoproterozoic Snowball Earth: A Climate Disaster Triggered by the evolution of Oxygenic Photosynthesis." *Proceedings of the National Academy of Sciences* 102, no. 32 (2005): 11131–11136. https://doi.org/10.1073/pnas.0504878102.

Olson, Kenneth R., and Karl D. Straub. "The Role of Hydrogen Sulfide in Evolution and the Evolution of Hydrogen Sulfide in Metabolism and Signaling." *Physiology* 31, no. 1 (2016): 60–72. https://doi.org/10.1152/physiol.00024.2015.

Orme, A. R. "Early Earth." ScienceDirect. Last modified April 17, 2022. https://www.sciencedirect.com/topics/earth-and-planetary-sciences/early-earth.

Plain, Charlie. "NASA Finds Evidence Two Early Planets Collided to Form Moon." NASA. Last modified September 17, 2020. https://www.nasa.gov/feature/nasa-finds-evidence-two-early-planets-collided-to-form-moon.

Smithsonian. "A Collection of Cambrian Fossils." Last modified November 1, 2022, https://ocean.si.edu/through-time/ancient-seas/collection-cambrian-fossils.

Wolpert, Stuart. "Moon Was Produced by a Head-on Collision between Earth and a Forming Planet." UCLA. January 28, 2016. https://newsroom.ucla.edu/releases/moon-was-produced-by-a-head-on-collision-between-earth-and-a-forming-planet.

Zahnle, Kevin, Laura Schaefer, and Bruce Fegley. "Earth's Earliest Atmospheres." *Cold Spring Harbor Perspectives in Biology* 2, no. 10 (2010): a004895. https://doi.org/10.1101/cshperspect.a004895.

REASON 26

Fernández, Lucia Ayala, Carsten Wiedemann, and Vitali Braun. "Analysis of Space Launch Vehicle Failures and Post-Mission Disposal Statistics." *Aerotecnica Missili & Spazio* 2022. https://doi.org/10.1007/s424 96-022-00118-5.

Gant, Phylindia, and Amy J. Williams. "Could People Breathe the Air on Mars?" The Conversation. May 16, 2022. https://theconversation.com /could-people-breathe-the-air-on-mars-180504.

IMDb. "A Trip to Mars." Last modified October 28, 2022. https://www.imdb .com/title/tt0008100/.

James, Chris. "How to Get People from Earth to Mars and Safely Back Again." The Conversation. December 20, 2020. https://theconversation.com /how-to-get-people-from-earth-to-mars-and-safely-back-again-150167.

NASA. "How Long Would a Trip to Mars Take?" Last modified April 4, 2022. https://image.gsfc.nasa.gov/poetry/venus/q2811.html.

NASA. "Mars Climate Orbiter." NASA Solar System Exploration. Last modified July 25, 2019. https://solarsystem.nasa.gov/missions/mars-climate -orbiter/in-depth/.

NASA. "Mars Exploration Program." NASA Science. Last modified January 13, 2022. https://mars.nasa.gov/.

Newman, Dava J. "Life in Extreme Environments: How Will Humans Perform on Mars?" *Gravitational and Space Biology* 12, no. 2 (2007): 35. http:// gravitationalandspaceresearch.org/index.php/journal/article/view/243/242.

Ojha, Lujendra, Mary Beth Wilhelm, Scott L. Murchie, Alfred S. McEwen, James J. Wray, Jennifer Hanley, Marion Massé, et al. "Spectral Evidence for Hydrated Salts in Recurring Slope Lineae on Mars." *Nature Geoscience* 8 (2015): 829–832. https://doi.org/10.1038/ngeo2546.

Sheetz, Michael. 2020. "Elon Musk Is 'Highly Confident' SpaceX Will Land Humans on Mars by 2026." CNBC. December 1, 2020. https:// www.cnbc.com/2020/12/01/elon-musk-highly-confident-spacex-will -land-humans-on-mars-by-2026.html.

Welch, Richard, Daniel Limonadi, and Robert Manning. "Systems Engineering the Curiosity Rover: A Retrospective." *2013 8th International Conference on System of Systems Engineering* (2013): 70–75. https://doi.org/10.1109 /sysose.2013.6575245.

Yamashita, Masamichi, Yoji Ishikawa, Yoshiaki Kitaya, Eiji Goto, Mayumi Arai, Hirofumi Hashimoto, Kaori Tomita-Yokotani, et al. "An Overview of Challenges in Modeling Heat and Mass Transfer for Living on Mars."

Annals of the New York Academy of Sciences 1077, no. 1 (2006): 232–243, https://doi.org/10.1196/annals.1362.012.

REASON 27

Alekhova, T. A., L. M. Zakharchuk, N. Y. Tatarinova, V. V. Kadnikov, A. V. Mardanov, N. V. Ravin, and K. G. Skryabin. "Diversity of Bacteria of the Genus *Bacillus* on Board of International Space Station." *Doklady Biochemistry and Biophysics* 465, no. 1 (2015): 347–350. https://doi.org/10.1134/s1607672915060010.

Bishop, Forrest. "Open Air Space Habitats." Institute of Atomic-Scale Engineering, 1997. Last modified July 25, 2018. https://www.iase.cc/openair.htm.

Demontis, Gian C., Marco M. Germani, Enrico G. Caiani, Ivana Barravecchia, Claudio Passino, and Debora Angeloni. "Human Pathophysiological Adaptations to the Space Environment." *Frontiers in Physiology* 8 (2017): 547. https://doi.org/10.3389/fphys.2017.00547.

Ferl, Jinny, L. Hewes, D. Cadogan, David Graziosi, and Keith Splawn. "System Considerations for an Exploration Spacesuit Upper Torso Architecture." *SAE Technical Paper* (2006): https://doi.org/10.4271/2006-01-2141.

Fisher, Nick. "Space Science 2001: Some Problems with Artificial Gravity." *Physics Education* 36, no. 3 (2001): 193–201. https://doi.org/10.1088/0031-9120/36/3/303.

Starr, Michelle. "What Happens to the Unprotected Human Body in Space?" CNET. July 27, 2014. https://www.cnet.com/culture/what-happens-to-the-unprotected-human-body-in-space/.

Vernikos, J., and V. S. Schneider. "Space, Gravity and the Physiology of Aging: Parallel or Convergent Disciplines? A Mini-review." *Gerontology* 56, no. 2 (2010): 157–166. https://doi.org/10.1159/000252852.

Williams, David, Andre Kuipers, Chiaki Mukai, and Robert Thirsk. "Acclimation During Space Flight: Effects on Human Physiology." *Canadian Medical Association Journal* 180, no. 13 (2009): 1317–1323. https://doi.org/10.1503/cmaj.090628.

Wilson, J. W., B. M. Anderson, F. A. Cucinotta, J. Ware, and C. J. Zeitlin. "Spacesuit Radiation Shield Design Methods." *SAE Transactions* 115, no. 1 (2006): 277–293. https://www.jstor.org/stable/44657683.

REASON 28

Baker, Daniel N., Roberta Balstad, Michael Bodeau, and Eugene Cameron. *Severe Space Weather Events: Understanding Societal and Economic Impacts:*

A Workshop Report. Washington, D.C.: National Academies Press, 2008. https://doi.org/10.17226/12507.

Chapman, S. C., R. B. Horne, and N. W. Watkins. "Using the aa Index over the Last 14 Solar Cycles to Characterize Extreme Geomagnetic Activity." *Geophysical Research Letters* 47, no. 3 (2020): e2019GL086524. https://doi.org/10.1029/2019GL086524.

Dyer, C. S., and P. R. Truscott. "Cosmic Radiation Effects on Avionics." *Microprocessors and Microsystems* 22, no. 8 (1999): 477–483. https://doi.org/10.1016/S0141-9331(98)00106-9.

Fujita, Moe, Tatsuhiko Sato, Susumu Saito, and Yosuke Yamashiki. "Probabilistic Risk Assessment of Solar Particle Events Considering the Cost of Countermeasures to Reduce the Aviation Radiation Dose." *Scientific Reports* 11, no. 1 (2021): 17091. https://doi.org/10.1038/s41598-021-95235-9.

Giegengack, Robert. "The Carrington Coronal Mass Ejection of 1859." *Proceedings of the American Philosophical Society* 159, no. 4 (2015): 421–433. https://www.jstor.org/stable/26159195.

Hellerstedt, John. "February 2021 Winter Storm-Related Deaths—Texas." Texas Department of State Health Services. December 31, 2021. Last modified May 13, 2022. https://www.dshs.texas.gov/news/updates/SMOC_FebWinterStorm_MortalitySurvReport_12-30-21.pdf.

Kappenman, John, and William Radasky. "Geomagnetic Field Impacts on Ground Systems." In *Space Weather Effects and Applications*, edited by Anthea J. Coster, Philip J. Erikson, Louis J. Lanzerotti, Yongliang Zhang, and Larry J. Paxton, 183–213. Washington, D.C.: American Geophysical Union, 2021. https://doi.org/10.1002/9781119815570.ch9.

Schieb, Pierre-Alain, and Anita Gibson. "Geomagnetic Storms." Office of Risk Management and Analysis, United States Department of Homeland Security. January 14, 2011. Last modified April 25, 2021. http://www.oecd.org/gov/risk/46891645.pdf.

Shibata, Kazunari, Hiroaki Isobe, Andrew Hillier, Arnab Rai Choudhuri, Hiroyuki Maehara, Takako T. Ishii, Takuya Shibayama, et al. "Can Superflares Occur on Our Sun?" *Publications of the Astronomical Society of Japan* 65, no. 3 (2013): 49. https://doi.org/10.1093/pasj/65.3.49.

Yates, Athol. "Death Modes from a Loss of Energy Infrastructure Continuity in a Community Setting." *Journal of Homeland Security and Emergency Management* 10, no. 2 (2013): 587–608. https://doi.org/10.1515/jhsem-2012-0048.

REASON 29

Costello, Mark J., Robert M. May, and Nigel E. Stork. "Can We Name Earth's Species Before They Go Extinct?" *Science* 339, no. 6118 (2013): 413–416. https://doi.org/10.1126/science.1230318.

Lasher, Larry, and John Dyer. "Pioneer Missions." In *Encyclopedia of Astronomy & Astrophysics*, edited by P. Murdin, 14. Boca Raton, FL: CRC Press, 2000.

Lopez, Hugo. "The Protection of Cultural Heritage Sites on the Moon: The Poo Bags Paradox." In *Protection of Cultural Heritage Sites on the Moon*, edited by Annette Froehlich, 131–143. Cham, Switzerland: Springer, 2020. https://doi.org/10.1007/978-3-030-38403-6_11.

Mayer, Larry, Martin Jakobsson, Graham Allen, Boris Dorschel, Robin Falconer, Vicki Ferrini, Geoffrey Lamarche, et al. "The Nippon Foundation—GEBCO Seabed 2030 Project: The Quest to See the World's Oceans Completely Mapped by 2030." *Geosciences* 8, no. 2 (2018): 63. https://doi.org/10.3390/geosciences8020063.

Schwieterman, Edward W., Nancy Y. Kiang, Mary N. Parenteau, Chester E. Harman, Shiladitya DasSarma, Thresa M. Fisher, Giad N. Arney, et al. "Exoplanet Biosignatures: A Review of Remotely Detectable Signs of Life." *Astrobiology* 18, no. 6 (2018): 663–708. https://doi.org/10.1089/ast.2017.1729.

Socas-Navarro, Hector, Jacob Haqq-Misra, Jason T. Wright, Ravi Kopparapu, James Benford, and Ross Davis. "Concepts for Future Missions to Search for Technosignatures." *Acta Astronautica* 182 (2021): 446–453. https://doi.org/10.1016/j.actaastro.2021.02.029.

Sparks, W. B., K. P. Hand, M. A. McGrath, E. Bergeron, M. Cracraft, and S. E. Deustua. "Probing for Evidence of Plumes on Europa with HST/STIS." *The Astrophysical Journal* 829, no. 2 (2016): 121. https://doi.org/10.3847/0004-637X/829/2/121.

REASON 30

Cox, T. J., and Abraham Loeb. "The Collision between the Milky Way and Andromeda." *Monthly Notices of the Royal Astronomical Society* 386, no. 1 (2008): 461–474. https://doi.org/10.1111/j.1365-2966.2008.13048.x.

Dar, Arnon, Ari Laor, and Nir J. Shaviv. "Life Extinctions by Cosmic Ray Jets." *Physical Review Letters* 80, no. 26: 5813–5816. https://doi.org/10.1103/PhysRevLett.80.5813.

Elvis, Martin. "A Structure for Quasars." *The Astrophysical Journal* 545, no. 1 (2000): 63–76. https://doi.org/10.1086/317778.

Giommi, Paolo. "Multi-frequency, Multi-messenger Astrophysics with Swift.

The Case of blazars." *Journal of High Energy Astrophysics* 7 (2015): 173–179. https://doi.org/10.1016/j.jheap.2015.06.001.

Hardcastle, Torrie. "Texas Town Named Most Boring city in America." *Houston Chronicle,* September 29, 2014. https://www.mysanantonio.com/homes/article/Texas-town-named-most-boring-city-in-America-5789043.php#photo-5300446.

NASA. "Spitzer Captures Messier 87." NASA Jet Propulsion Laboratory. April 25, 2019. https://www.jpl.nasa.gov/images/pia23122-spitzer-captures-messier-87.

REASON 31

Al Kharusi, S., S. Y. Benzvi, J. S. Bobowski, W. Bonivento, V. Brdar, T. Brunnar, E. Caden, et al. "SNEWS 2.0: A Next-Generation Supernova Early Warning System for Multi-messenger Astronomy." *New Journal of Physics* 23 (2021): 031201. https://doi.org/10.1088/1367-2630/abde33.

Cannon, Annie Jump, and Edward Charles Pickering. "Classification of 1,688 Southern Stars by Means of Their Spectra." *Annals of the Astronomical Observatory of Harvard College* 56, no. 5 (1912): 115. https://ui.adsabs.harvard.edu/abs/1912AnHar..56..115C.

Meynet, G., L. Haemmerlé, S. Ekström, C. Georgy, J. Groh, and A. Maeder. "The Past and Future Evolution of a Star Like Betelgeuse." *EAS Publications Series* 60 (2013): 17–28. https://doi.org/10.1051/eas/1360002.

Tokovinin, A. A. "MSC—A Catalogue of Physical Multiple Stars." *Astronomy and Astrophysics Supplement Series* 124, no. 1 (1997): 75. https://doi.org/10.1051/aas:1997181.

REASON 32

Alexander, Conel M.O'D, and George W. Wetherill. "Meteor and Meteoroid." Encyclopedia Britannica. Last modified September 27, 2022. https://www.britannica.com/science/meteor.

Bostrom, Nick. "Existential Risks: Analyzing Human Extinction Scenarios and Related Hazards." *Journal of Evolution and Technology* 9, no. 1 (2002). https://nickbostrom.com/existential/risks.

Dinosaurs: A Children's Encyclopedia, 2nd ed. London: Dorling Kindersley Limited, 2019.

IMDb. "The Violent Past." IMDb. Last modified October 28, 2022. https://www.imdb.com/title/tt2255057/.

NASA. "Discovery Statistics." Center for Near Earth Object Studies. Last modified November 3, 2022. https://cneos.jpl.nasa.gov/stats/site_140.html.

Pravec, Petr, Alan W. Harris, Peter Kušnirák, Adrián Galád, and Kamil Hornoch. "Absolute Magnitudes of Asteroids and a Revision of Asteroid Albedo Estimates from WISE Thermal Observations." *Icarus* 221, no. 1 (2012): 365–387, https://doi.org/10.1016/j.icarus.2012.07.026.

REASON 33

Hoffman, Ashley. "Paul Rudd Reveals the Secret Behind Being an Ageless Baby-Faced Adult Man." *TIME*. March 25, 2019. https://time.com/5558046/paul-rudd-age-clueless/.

Schröder, K.-P., and Robert Connon Smith. "Distant Future of the Sun and Earth Revisited." *Monthly Notices of the Royal Astronomical Society* 386, no. 1 (2008): 155–163. https://doi.org/10.1111/j.1365-2966.2008.13022.x.

Wolf, E. T., and O. B. Toon. "Delayed Onset of Runaway and Moist Greenhouse Climates for Earth." *Geophysical Research Letters* 41 (2013): 167–172. https://doi.org/10.1002/2013GL058376.

REASON 34

Novikov, Igor. "Black Holes." In *Stellar Remnants*. Berlin: Springer, 1997, https://doi.org/10.1007/3-540-31628-0_3.

Pinochet, Jorge. "The Little Robot, Black Holes, and Spaghettification." *Physics Education* 57, no. 4 (2022): 045008. https://doi.org/10.1088/1361-6552/ac5727.

The Event Horizon Telescope Collaboration, Kazunori Akiyama, Antxon Alberdi, Walter Alef, Keiichi Asada, Rebecca Azulay, Anne-Kathrin Baczko, et al. "First M87 Event Horizon Telescope Results. I. The Shadow of the Supermassive Black Hole." *The Astrophysical Journal Letters* 875, no. 1 (2019): L1. https://doi.org/10.3847/2041-8213/ab0ec7.

REASON 35

Carnot, Sadi. *Reflections on the Motive Power of Heat*, edited by R. Henry Thurston. New York: J. Wiley & Sons, 1890. https://openlibrary.org/books/OL14037447M.

Thomson, Sir William. "On the Age of the Sun's Heat." *Macmillan's Magazine* 5 (1862): 388–392.

REASON 36

Caldwell, Robert R., Marc Kamionkowski, and Nevin N. Weinberg. "Phantom Energy: Dark Energy with w<−1 Causes a Cosmic Doomsday." *Physical Review Letters* 91, no 7 (2003): 071301. https://doi.org/10.1103 /PhysRevLett.91.071301.

Eicher, David J. "How Many Galaxies Are There? Astronomers Are Revealing the Enormity of the Universe." *Discover*, May 19, 2020. https:// www.discovermagazine.com/the-sciences/how-many-galaxies-are -there-astronomers-are-revealing-the-enormity-of-the.

REASON 37

Madsen, Jes. "Physics and Astrophysics of Strange Quark Matter." In *Hadrons in Dense Matter and Hadrosynthesis*, edited by J. Cleymans, H. B. Geyer, and F. G. Scholz. Berlin: Springer https://doi.org/10.1007/BFb0107314.

Weber, F. "Strange Quark Matter and Compact Stars." *Progress in Particle and Nuclear Physics*. 54, no. 1 (2005): 193. https://doi.org/10.1016 /j.ppnp.2004.07.001.

REASON 38

Callan, Curtis G., and Sidney Coleman. "Fate of the False Vacuum. II. First Quantum corrections." *Physical Review D* 16, no. 6 (1977): 1762. https:// doi.org/10.1103/PhysRevD.16.1762.

Mack, Katie. "Vacuum Decay: The Ultimate Catastrophe." *Cosmos*, September 14, 2015. https://cosmosmagazine.com/science/physics/vacuum-decay -the-ultimate-catastrophe/.

REASON 39

Bostrom, Nick. "Are We Living in a Computer Simulation?" *Philosophical Quarterly* 53, no. 211 (2003): 243–255. https://doi.org/10.1111/1467 -9213.00309.

Kelly, Stephen. "The Matrix: Are We Living in a Simulation?" BBC Science Focus. January 18, 2022. https://www.sciencefocus.com/future-technology /the-matrix-simulation/.

REASON 40

Carr, Bernard J., and Steven B. Giddings. "Quantum Black Holes." *Scientific American* 292, no. 5 (2022): 48–55. https://doi.org/10.1038/scientific american0505-48.

Hawking, Stephen W. "Gravitationally Collapsed Objects of Very Low Mass." *Monthly Notices of the Royal Astronomical Society* 152, no. 1 (1971): 75. https://doi.org/10.1093/mnras/152.1.75.

Sokol, Joshua. "Physicists Argue That Black Holes from the Big Bang Could Be the Dark Matter." *Quanta Magazine.* September 23, 2020. https://www.quantamagazine.org/black-holes-from-the-big-bang-could-be-the-dark-matter-20200923/.

REASON 41

Randall, Lisa. *Dark Matter and the Dinosaurs: The Astounding Interconnectedness of the Universe.* London: Random House, 2016.

REASON 42

Aghanim, N., Y. Akrami, M. Ashdown, J. Aumont, C. Baccigalupi, M. Ballardini, A. J. Banday, et al. "Planck 2018 Results: VI. Cosmological Parameters." *Astronomy & Astrophysics* 641 (2020): A6. https://doi.org/10.1051/0004–6361/201833910.

ABOUT THE AUTHORS

Chris Ferrie is an associate professor at the University of Technology Sydney in Australia, where he researches and lectures on quantum physics, computation, and engineering. He is the author of *Quantum Bullsh*t: How to Ruin Your Life with Quantum Physics* and also happens to be the number one bestselling science author for kids. Though those books have decidedly fewer f-bombs.

Wade David Fairclough is an Australian author, illustrator, and educator. His work has been nominated for multiple awards, such as the Australian Podcast of the Year—Science and Technology, and has also charted on iTunes Comedy Podcast in Australia, Europe, and North America. He is the coauthor of *Pranklab: 25 Hilarious Scientific Practical Jokes for Kids.*

Byrne LaGinestra is a science communicator known for his unique approach to education. He has a diverse body of work from textbooks and children's books to public presentations and podcasts. Based in Australia, his work has received widespread recognition both nationally and internationally. He has helped teachers, students, and enthusiasts alike learn to love science. Until now.

To learn more about ways the universe is out to get you (and other science-y things), visit www.42reasonstohatetheuniverse .com.